C0-ATR-896

ENERGY AND HUMAN WELFARE—A CRITICAL ANALYSIS

VOLUME I
The Social Costs of Power Production

ENERGY AND HUMAN WELFARE—A CRITICAL ANALYSIS

A Selection of Papers on the Social, Technological, and Environmental Problems of Electric Power Consumption

---

Edited by
BARRY COMMONER
HOWARD BOKSENBAUM
MICHAEL CORR

Prepared for the Electric Power Task Force of the Scientists' Institute for Public Information and the Power Study Group of the American Association for the Advancement of Science Committee on Environmental Alterations

VOLUME I

The Social Costs of Power Production

MACMILLAN INFORMATION
A Division of Macmillan Publishing Co., Inc.
New York

Macmillan Information
A Division of Macmillan Publishing Co., Inc.
866 Third Avenue, New York, N.Y. 10022

Collier-Macmillan Canada Ltd.

Library of Congress Catalog Card Number: 75-8986

Printed in the United States of America

printing number

1 2 3 4 5 6 7 8 9 10

Library of Congress Cataloging-in Publication Data
Main entry under title:

The Social costs of power production.

(Energy and human welfare ; v. 1)
Includes bibliographies and index.
1. Electric power-plants--Environmental aspects --Addresses, essays, lectures. I. Commoner, Barry, 1917- II. Boksenbaum, Howard. III. Corr, Michael. IV. Scientists' Institute for Public Information. Electric Power Task Force. V. American Association for the Advancement of Science. Committee on Environmental Alterations. Power Study Group. VI. Series.
TD195.E4S6 363.6 75-8986
ISBN 0-02-468420-1

Contents

STATEMENT OF THE BOARD OF DIRECTORS OF THE AMERICAN ASSOCIATION FOR THE ADVANCEMENT OF SCIENCE

The following expression of the Board's Statement of Policy concerning AAAS Committee Activities and Reports: "Responsibility for statements of fact and expressions of opinion contained in this report rests with the committee that prepared it.* The AAAS Board of Directors, in accordance with Association policy and without passing judgment on the views expressed, has approved its publication as a contribution to the discussion of an important issue."

*Relevant to this statement of policy, the Committee on Environmental Alterations commissioned the Power Study Working Group to select the authors for the preparation of this report. The contributors to the report are responsible for the statements of fact and expressions of opinion in their respective papers.

Members of the AAAS-CEA who participated in organizing these volumes and their affiliation at that time:

Barry Commoner, Washington University
Donald Aitken, San Jose State College
F. Herbert Borman, Yale University
Theodore Byerly, U.S. Department of Agriculture
William Capron, Harvard University
H. Jack Geiger, SUNY at Stony Brook
Oscar Harkavy, Ford Foundation
Walter Modell, College of Medicine, Cornell University
Arthur Squires, City University of New York

STATEMENT OF THE COMMITTEE ON ENVIRONMENTAL ALTERATIONS

It has become clear that the majority of the people of the United States are working to stem the escalation of environmental problems resulting from the incessant expansion of industrial technology. It is this Committee's purpose to provide the public with an assessment of the environmental consequences of our technology.

Energy-related environmental problems and their associated social issues are particularly pressing since the energy industry and, in particular, the electric power industry, appear to be a major force in our economy. Lead times for major decisions in the electric power industry are of the order of 10 to 25 years, which makes it difficult for the public, the news media, and policy makers to engage in timely, well-informed debate over the crucial environmental and social problems created by increasing energy consumption.

To help fill the need for documentation on these problems, the Committee on Environmental Alterations, in collaboration with the Scientists' Institute for Public Information, established a Task Force to prepare a report on the environmental impact of power production and possible means of alleviating adverse effects. One result of the work of this Task Force has been the preparation of these three volumes. In the view of the Committee, this is a useful contribution to our understanding of this complex and urgent problem.

We recognize that each of the authors of the separate papers that make up the main part of the report have strong, well-developed views on their subjects which may not be shared by their colleagues. This is to be expected in a field which is still poorly understood. Where sharp differences of approach to a given problem exist, the report includes, if only in brief form, some material indicative of the divergence in views.

The authors of each article take full responsibility for their contributions and, on its part, the Committee on Environmental Alterations hopes that members of the scientific community and of the public at large will find this compendium a useful contribution to an increasingly urgent subject.

The Committee wishes to express its particular appreciation to Professor Dean Abrahamson and to the Task Force members for their efforts in providing this valuable document.

Acknowledgements

The editors wish to express their gratitude to all of the authors whose cooperation was so important to the production of this volume: Special thanks go to Artis Bernard for her help in resolving questions of style. Equally important was the efficient and expert help of those who contributed to the preparation of the manuscript: the artwork was done by Dru Lipsitz and Lurline Hogsett; typing and maintenance of mail communications were handled by Gladys Yandell, Peggy Whitton, Amy Papian, Lenore Harris, Jane Murdock, Sandy Marshall, Leslie Rutlin, Pieriette Murray, Cynthia Glastris and Julie Love.

We would also like to thank J. Klarmann, K. H. Lusczynski, Michael Friedlander and Caleb A. Smith for their assistance.

In addition, we wish to acknowledge the cooperation of various publishers in granting us permission to reproduce those papers that have also appeared elsewhere. Certain chapters have appeared as articles in *Environment* magazine, and are copyrighted in the indicated year by the Scientists' Institute for Public Information:

"Acid Rain," March 1972
"A Small Oil Spill," March 1971
"Radiation: The Invisible Casualties," April 1970

Chapter 1 was accepted for publication by *Environmental Affairs* as "Thermal Pollution and Its Control," Fall 1972.

Chapter 5 was adapted from *Research Journal*, 51, Ag. Exp, Sta., University of Wyoming, Laramie, September 1971, with permission of the authors.

Chapter 8 appeared as "Radioactivity from Fossil-Fuel and Nuclear Power Plants" in *Proceedings of the Conference on Atomic Power and the Environment,* International Atomic Energy Agency, SM-146/19.

Chapter 11 appeared previously in a revised form in *Technology Review,* March/April 1972.

Preface

One of the remarkable things about the use of power in the United States and other industrialized countries is that it has grown *exponentially*—that the rate of growth has itself been increasing. This suggests that the growth process is self-accelerating—that the very use of power seems to encourage the use of more power.

Such self-propelled processes are well known in science. A favorite example is the growth of bacteria in a test tube of nutrient. As a few starting cells divide they double in number, let us say in the course of an hour; one hour later, their offspring have again divided and the number of new cells produced per hour is doubled; and so on. Growth begets growth and the outcome of this mathematical abstraction is the familiar, ever-rising exponential curve.

However, anyone who has grown bacteria in a test tube—or for that matter rabbits in a hutch, or cattle in a pasture—knows, this is only part of the story. Growth always slows down and comes to a halt as the food available in the test tube (or in the pasture) becomes insufficient to support the growth of the new offspring. When the abstract idea that growth begets growth is translated into reality, constraints appear which slow down and eventually bring the exponential process to a halt.

To generalize a bit we can say that the existence of an exponential growth curve is a sign that the driving process has not *yet* encountered the opposing forces that sooner or later will overcome it.

When we look at the still-accelerating growth in the use of power in the United States we need to ask, then, what opposing processes will eventually slow down the rate of growth and bring it to a halt? One constraint is, of course, the availability of fuel. If, for example, the accelerating use of power is based on the burning of fossil fuel—coal, oil and gas—as it now is very largely in the United States, then at some point the fuel will be used up and the rate of use will not only stop accelerating but eventually fall to zero.

The second constraint appears to be both as pressing and as unavoidable as the finitude of fuel resources. This is the social cost of environmental degradation caused by the generation of power. The motivation for the production of power is social—that it yields goods and services that society values. However, as the social costs of producing power—environmental degradation—rise they can reduce the motivation to produce power, if the environmental cost is judged to be too large to justify the benefits of power.

These social costs of power production include dangers to human health from pollution emitted by power plants, destruction of valued natural resources such as clean air and clear streams and, indeed, of the fuel itself (and the resultant hindrance to productive processes that require these resources), and the blighting of scenic places with power lines and generating plants. These social costs are the subject of this volume.

Different kinds of power plants produce different kinds of environmental damage. However, there is one environmental pollutant that is common to *all* power plants: heat. Physical principles tell us that no power plant can be 100% efficient, that some of the energy in the fuel is not converted to power, but is lost to the environment as heat. Present day fossil fuel plants return as electricity only up to 40% of the heat value of the fuel they consume; nuclear plants convert about 33% of the heat of fission to electric power. The remaining energy, 60% or more, is waste heat nearly all of which is released to the environment as thermal pollution. Thermal pollution by power plants is inescapable. In order to extract energy from the fuel, the power plant must dispose of the waste heat. Given the laws of physics this means that the heat must be transferred to some cooler part of the environment—usually to the cool waters of a river or lake. That this interaction between power production and the environment is a serious constraint on further power production is evident from the following fact: it is estimated that the new power plants needed to meet demand projected for the next 20 years alone could require for their cooling the total annual runoff of water from the 48 continguous United States.

Does this mean that nothing can be done about waste heat from power generation? Chapter 1, which discusses this problem, provides an illuminating answer: The heat emitted by a power plant need not be entirely wasted, provided that we expand our vision of the purpose of a power plant. If a power plant can be conveniently and safely located near homes then the heat that it emits, although at too low a temperature to be useful for power generation, can be used to warm the homes, There is a lesson here: We need to learn how to think about technology in *holistic terms,* to see that a power plant when viewed as a whole produces more than the power that motivates its design; to realize that one of its products, waste heat, can be put to use if the design encompasses not only the production of power but the plant's relationship to its setting.

The environmental effects of waste heat are not yet well understood. As the problem grows, questions about the urban "heat island" effect, about the action of waste heat on aquatic and climatic systems, and about the effect of increased ambient or local temperatures on the complex of air and water pollutants will have to be investigated. Global implications as well as the local effects will have to be taken into account. Working from energy need projections, some scientists predict a measurable increase in the surface temperature of the earth by the year 2020. However, the interrelationship between heat released by the burning of fuels and other processes that affect the earth's temperature is only poorly known. We are, in fact, a long way from fully evaluating the social costs of power production that are caused by waste heat.

Power plants produce more than power and waste heat: they also emit into the environment substances that degrade its quality. Power plants fired by fossil fuels stand out as primary contributors to the massive air pollution that now envelopes all industrialized regions of the world. Air pollution lends itself more easily than thermal pollution to rough estimations of social costs. One estimate is that air pollution from electric power plants caused over $8 billion in health and property damage in 1972—over $40 per year for every man, woman and child in the United States. This total is made up of costs which range from increased bills for laundry, for repair of corrosion-damaged buildings, to increased costs of medical care. But the figure is only a rough one because many of the effects of air pollution on health are difficult to estimate, for example, because the synergistic interactions of various pollutants are still poorly understood (this is discussed in Chapter 2).

Plants as well as people are damaged by air pollution from power plants. Photosynthesis, and therefore plant growth, is retarded by low levels of such pollutants as ozone, thus threatening the productivity of forest and crops. In California vegetable crops are lost with increasing frequency due to the effects of air pollution. Rains acidified by sulphur dioxide, a major pollutant emitted by fossil fuel power plants, can cause considerable damage to vegetation, the soil and buildings (Chapter 3). This has become a major problem in Northern Europe and has now begun to affect areas of the United States. In many places—Venice, Florence and other Italian cities are classic examples—corrosive pollutants and the synergistic action of many air pollutants have begun to destroy sculptures which had withstood centuries of weathering until the advent of modern pollutants.

Although pollution is the most visible social cost of power production from fossil fuels, it is not the only one. Major social costs are incurred in the extraction of coal and in the effects of coal mining on the health and safety of miners (see Chapters 4 and 5). Much less is known about the corresponding effects of oil and gas wells, except for a few reports about the ecological effects of oil lost from underwater wells, such as those near Santa Barbara, California. Here, too, is an area in which the social costs of powder production are but poorly known.

Power production also incurs important social costs through the transportation of fuel. The best known example of such costs—oil spilled from tankers—is taken up in Chapter 6. This problem is likely to become increasingly serious as more and more of the ocean transport of oil is taken over by huge supertankers—the largest and least manageable ships in history. The portentous story of the supertankers is brilliantly told in a recent book by Noel Mostert entitled *Supership*.

The most celebrated example of such social costs is the Alaska Pipeline, which provides us with some illuminating, if discouraging lessons. When a large oil field was discovered on the North Slope of Alaska, the companies that had bought the right to extract the oil from the State of Alaska at once proposed to lay a pipeline from the area to the nearest frost-free harbor—a distance of some 800 miles. Their proposal ignored the special environmental features of the land that the pipeline would travel, in particular that most of it was in *permafrost*—permanently frozen subsoil that would be locally thawed by the heated pipeline. But through the legal efforts of conservation organizations, and especially with the passage of the National Environmental Policy Act (NEPA) in 1970, the government was required to evaluate and make public the full environmental impact of the pipeline. The resultant Environmental Impact Statement—even though it dealt with a new pipeline design highly modified to reduce its impact on the permafrost—convincingly showed that the pipeline would in fact have an appreciable effect on the tundra environment. Nevertheless, in 1973 under the pressure of a supposed shortage of oil in the U.S., Congress enacted legislation that empowered the construction of the pipeline. This is sometimes regarded as an "environmental defeat" and evidence that the provision of NEPA for environmental impact statements is futile. Let us recall, however, that there are no absolute, objective or "scientific" ways to balance the benefits of a given social action against its costs, for this is always a matter of judgement that necessarily reflects public mores and political outlook. What NEPA has accomplished in the case of the Alaska Pipeline is not to be minimized: It forced us to make this judgement with our eyes open to its ecological consequences, so that if, after the pipeline is in operation, the social costs appear to outweigh the benefits, we will be aware of what this mistake has cost us. Such an awareness may make us wiser at the next opportunity for a mistake.

Other social costs not considered in this volume include opportunity costs of land used for the construction of power plants and transmission lines; aesthetic effects of power plants and transmission lines; and the political and economic ramifications of possibly being forced to rely upon foreign fuel supplies. Following the impact of the Arab oil embargo of 1973, the importance of this last factor hardly needs to be argued.

It is not enough to determine that there are social costs associated with the production of electric power. It is equally important to find out who bears the costs, and who gains from the benefits. A telling instance is provided by the Four Corners power complex. Six power plants are planned for completion by 1977 in the Four Corners Area at the convergence of the boundaries of Arizona, Colorado, New Mexico and Utah. The plants will be fueled with coal stripped from rich nearby deposits, including the Black Mesa. Power produced at Four Corners will supply electricity to Phoenix, Tucson, Las Vegas, and Southern California, all so far away from the power plants as to suffer none of the costs of cultural dislocation, strip mining, and pollution that accompany the construction and operation of the plants. By 1977, the Four Corners complex will be consuming 130,000 tons of coal daily. The social costs—intense air pollution and a highly disrupted landscape—will be borne by the residents of the Four Corners area, Hopi and Navajo Indians who will neither need nor receive much of the plants' power output.

Similarly inequitable distribution of costs and benefits from power production also occur in urban situations. Poor people often live in urban areas downwind from power plants and other industrial operations, but they use less electric power per family than the rich people who live upwind of the plants and breathe less of its pollutants.

Among the different ways of producing power, one—nuclear generating plants—is often cited as being relatively free of the environmental costs associated with the extraction and burning of fossil fuels. It is on these grounds that nuclear power is sometimes promoted as the "clean" alternative to fossil fuel. That the problem is not so simple is evident from the fact that it has been necessary to devote fully half the chapters of this book to it. Power generation from nuclear fission is still in its infancy. Although light water nuclear power reactors (LWRs) have been operating in the U.S. since 1957 (see Chapter 7) there are many uncertainties about the environmental impact of the whole system of nuclear power generation. It has been well established that uranium miners suffer an elevated incidence of cancer. Transportation of radioactive nuclear materials, such as fuels as spent fuel rods, involves a number of hazards; highly radioactive materials may leak from faulty containers or from containers damaged in traffic accidents. During storage and transport, reactor fuel is subject to theft. Stolen plutonium could, with relatively little difficulty, be made into a nuclear explosive (see Chapter 14)—a subject elaborated upon dramatically in a recent book by Mason Willrich and Theodore B. Taylor, *Nuclear Theft: Risks and Safeguards*.

Once inside an operating power plant, the safety of nuclear fuels still appears to be in doubt. Radioactivity emitted by modern nuclear power plants is expected to be greater than that emanating from modern fossil fuel plants. Nevertheless, it is believed that a reactor *could* be designed so that if it were well operated and diligently maintained it would pose less of a radiological threat to the environment than a plant using conventional fossil fuel (see Chapter 11). There is, however, serious controversy regarding the ability of present nuclear plants to continue operating properly. One aspect of this controversy, the efficacy of the emergency core cooling system, is discussed in Chapter 9. There are further dangers associated with spent fuel rod processing. These have been reduced during recent years, but further reduction is considered desirable (Chapter 13), and indeed in January, 1974, the AEC did in fact establish more stringent rules in this area. Finally, nuclear power plants accumulate highly radioactive wastes, which take many hundreds of years to decay. One of these wastes, plutonium 239, requires a period of time greater than the history of man on earth before it becomes innocuous. The cost of keeping these wastes quarantined throughout their toxic life must also be tallied among the social costs of nuclear power (Chapter 11). A new proposal for sealing wastes in the ice cap of Antartica, though not considered here, demands careful assessment.

Control of the dangers of nuclear power hinges upon a full understanding of the nature of the toxicity of radioactive materials—again a problem thrust upon us before we understand it. A description of the dangers of inhaling plutonium 239 (Chapter 12) illustrates the enormous complexity of this problem. For a number of years a fundamental question about the relation of radiation exposure to the resultant biological damage—whether there is a "threshold" dose, below which there is *no* effect—remained unresolved. It is worth noting that the papers represented by Chapters 12 and 13 themselves played an important role in recent acceptance of the linear hypothesis as opposed to the threshold hypothesis in predicting radiation damage to living organisms. We now know, therefore, that *every* added exposure to radiation, however slight, exacts an added social cost.

Perhaps more than any other form of power production nuclear power illustrates what may be the most far-reaching social cost of power production: The hazard to the processes of social judgement themselves of undertaking such huge and portentous activities before we are aware that they incur costs as well as yielding benefits. Consider just a few of the salient facts: The development of nuclear power plants, occurring *after* the considerable war-time experience with massive amounts of radioactivity, has given us an effective scientific base for appreciating the hazards of radiation to human health and to ecological systems. Nevertheless, it is fair to say that the regulations governing the operation of nuclear power plants have not taken this information seriously into account. Even now there is no acceptable means of isolating the enormous amounts of radioactive wastes that the industry already produces from potential contact with ecosystems and people. Moreover, nuclear power has preempted vast sums of money which could have otherwise been used to develop less hazardous and more reliable sources of energy such as solar energy. At the same time the U.S. government failed to develop an overall policy on the production and use of energy, usually excusing the omission on the grounds that all conceivable needs for energy in the future could be provided at minimal environmental cost by nuclear energy. What this amounts to is energy policy by default.

We now face the consequences of this "policy": As the economy is severely wrenched by the need to conserve energy, farmers, auto workers and citizens generally suffer from the chaos induced by oil shortages, whether real or contrived, and environmental legislation designed and enacted after numerous Congressional hearings is hastily dismembered and weakened in order to "solve" the energy crisis. Here we see the consequence of failing to examine the full meaning of power production—its social costs as well as its benefits—in good time. Once an expensive investment was made in the nuclear power industry it became difficult to reconsider its effects—not only on the environment, but also on the need for an overall national energy policy. As a result we now face the difficult choice between allowing the present, poorly understood chaotic situation to continue, or taking necessarily drastic steps to create a rational energy policy even if that should require, for example, a severe curtailment of plans for developing nuclear power in favor of, let us say, solar energy.

Thus we have come face to face with the difficult task of understading at what costs we have won the benefits that derive from the production and use of power. This volume is dedicated to serve that need, so that we can judge how to balance the costs of benefits.

May 29, 1974 **Barry Commoner**

CHRISTOPHER T. HILL

Thermal Pollution and Its Control

"Thermal pollution" is waste heat released to the environment as the unavoidable by-product of the generation of electricity in power plants. This chapter discusses the magnitude of the thermal pollution problem, some of the concerns raised about it, some of the technologies designed to control it, and some beneficial uses which have been suggested for waste heat.

Attention has been focused on waste heat in recent years due to the rapid growth of aggregate electrical power generation and due to the growth in size of individual power plants. The average generation station retired between 1962 and 1965 had a capacity of 22 MWe (Bregman, 1969), whereas plants of 600 MWe are common today and plants of 2000 MWe or more are being contemplated. The water required to cool the steam condensers of a 1000 MWe power plant is on the order of 800 to 1200 cfs, which is greater than one-half the water demand of Los Angeles, Chicago, or Metropolitan New York City (Jaske and Touhill, 1970). This cooling water requirement is a significant portion of the total flow of many rivers; e.g., the Connecticut River at Vernon, Vermont, has an average flow of 10,830 cfs (U.S. Dept. of Interior, 1961) and even the mighty Mississippi has an annual average flow of only 175,000 cfs at St. Louis (U.S. Dept. of Interior, 1962).

In absolute terms the Federal Power Commission (1969) estimates that total annual waste heat discharge from electric power plants will increase from 6×10^{15} BTU in 1969 to 20×10^{15} BTU in 1990. Jimeson and Adkins (1971) suggest that waste will grow from 5.3×10^{15} BTU in 1970 to 28.4×10^{15} BTU in 1990.

WASTE HEAT FROM ELECTRIC POWER PLANTS

An unavoidable accompaniment of the generation of electricity in steam power plants is the production of large amounts of waste heat. Modern fossil fuel plants are able to convert only about 40% of the energy released by burning coal, oil, or gas into electricity. Of the remaining 60%, about 45% is transferred from the low pressure steam to cooling water in the condenser and 15% is carried up the stack in the exhaust gas or is lost in the plant's mechanical systems. Due to lower operating temperature limits, the present generation of nuclear-fueled plants is usually designed to convert only 33% of the energy released by nuclear fission into electricity. Of the remainder, about 62% is transferred to cooling water in the condenser and 5% is lost to mechanical inefficiency (FPC, 1969; Jimeson and Adkins, 1971).

Traditionally, power plants have disposed of their waste heat by withdrawing water from a river or lake, passing it through the steam condenser, and returning it directly to the source. In this practice, known as "once-through" or "run-of-the-river" cooling, the cooling water is heated 10 to 30°F, with a 15 to 20°F rise being usual. It is desirable from the point of view of turbine design and plant efficiency to achieve the lowest possible condenser temperature, and therefore to have available cooling water at low temperatures and in large amounts.

The generating efficiency of a power plant is often described in terms of its "heat rate" which is the number of BTUs which must be released by burning or nuclear fission to produce one kilowatt-hour of electricity. If a plant were 100% efficient its heat rate would be 3413 BTU, the energetic equivalent of one kilowatt-hour. However, the average U.S. power plant had a heat rate of 10,300 BTU in 1969 (FPC, 1969) and this figure has not changed much in recent years. Some older plants still in service have heat rates of 20,000 BTU or higher while the best plant in 1969 achieved a heat rate of 8707 BTU (FPC, 1971). Since a growing fraction of power plants under construction in the U.S. are less efficient nuclear ones, we may expect that the national average heat rate will no longer decrease but will actually increase in the years to come.

WATER DEMAND FOR STEAM ELECTRIC GENERATION

Enormous quantities of cooling water are required today by the electric power industry. It is estimated that 50% of all water used in the U.S. is used by industry and that 80% of that amount is required for cooling electric power plants (FWPCA, 1968).

It is estimated that a total of 111,000 cfs of fresh water was withdrawn for power plant cooling in 1970 (Jimeson and Adkins, 1971). In 1980 this will rise to 153,000 cfs and to 301,000 cfs in 1990, assuming that the 1968 thermal criteria (FWPCA, 1968) are adequate to protect the quality of water bodies. In addition, a withdrawal of salt water at 46,000 cfs was required in 1970. In 1980 this will rise to 133,000 cfs and to 288,000 cfs in 1990 (Jimeson and Adkins, 1971). This figure of 301,000 cfs is estimated to be 1/6 of the average rate of run-off of U.S. rivers (FPC, 1969).

Although the same water may be used for cooling many times on its way to the ocean, it is clear that an upper limit of cooling water availability will one day be reached.

What happens to heat when it is returned to a cooling water source? It is known that a large portion of the heat is lost to the atmosphere by evaporation of water from the river or stream, whereas smaller fractions are lost to the surrounding air and stream bed by radiation, conduction and convection. While the exact distribution of heat loss among the various mechanisms depends upon local conditions, about 0.5 to 1.5 gallons of water are evaporated for every KWH of electricity generated. The exact amount evaporated depends upon power plant efficiency, ambient weather conditions, and the nature of the water source. Jimeson and Adkins (1969) have estimated that 1400 cfs of fresh water were evaporated for electric power generation in 1970, and that this figure will rise to 4300 in 1980 and 10,100 in 1990. If all plants expected to be in operation in 1990 used either cooling ponds, cooling towers, or a long ocean outfall, the evaporation rates could be as high as 6600 cfs in 1980 and 14,700 cfs in 1990.

ENVIRONMENTAL EFFECTS OF HEATED WATER DISCHARGE

The discharge of water at a temperature of 20 or 30°F above that which prevails in a watercourse can have severe effects on aquatic organisms (FWPCA, 1968; Krenkel and Parker, 1969; Eisenbud and Gleason, 1969). If the temperature change is sufficiently high, healthy adult organisms can be killed by thermal shock. Knowledge in this area is inadequate and much study of individual species and complete aquatic ecosystems is needed. However, it is known that, even if healthy adults can survive elevated temperatures, they may fail to reproduce or they may become more susceptible to disease. Furthermore, metabolic rates and oxygen demand of fish are high at higher temperatures, while the dissolved oxygen-carrying capability of water decreases as temperature increases. The net result may be more rapid growth and aging in some species. Heated river regions may also prove to be barriers to migration of some species to foraging or spawning grounds.

Fish are known to be able to adjust to slowly changing temperatures and to survive at temperature extremes which would be lethal if reached suddenly (FWPCA). Therefore, even though some species are able to adjust to heightened river temperatures near power plant discharges, rapid changes in rate of heat discharge caused by plant start-up or shut-down can be disastrous to them.

In general, the capacity of a river to assimilate wastes such as municipal sewage or agricultural run-off decreases with increasing water temperature; again due to the drop in oxygen saturation concentration (Krenkel and Parker, 1969). Increased water temperature can also lead to the replacement of desirable green algae with bluegreens, which, in combination with overstimulated plant growth, can lead to eutrophication, or premature aging of rivers and lakes.

Organisms of all kinds may be killed or damaged if entrained in the cooling water as it passes through pumps and ducts and is heated in the condenser tubes. Organisms which can survive the shock of suddenly increased temperature may experience mechanical damage due to abrasion or to fluid turbulence.

In 1968 a detailed set of recommendations for thermal discharge limits were issued by the National Technical Advisory Committee (FWPCA, 1968a). They are summarized in the Federal Water Pollution Control Administration's report (1968b) and include but are not limited to, recommendations for maximum water discharge temperatures which will raise the temperaturee of a stream no more than 5°F; of the cold, lower part of a lake no more than 3°F; and of estuaries no more than 4°F during the fall, winter and spring and no more than 1.5°F during the summer. Limits are also suggested for maximum temperatures for various species of fish, although no information on temperature ranges compatible with well-being is available for most species.

It should be noted, however, that the suggested limitations on temperatures are to apply outside a region of warmer water, called the "mixing zone," around the warm water discharge. The "mixing zone," which may be several hundred feet in extent, is the region in which the heated water is supposed to mix with the natural water and to be cooled. As a result, the mixing zone has come to be thought of as that portion of a watercourse allowed to be more than 5°F above ambient temperature. The size and shape of a mixing zone will depend on the design of the discharge canal or tube, or the nature of the watercourse, and on the relative flow rates of heated and background water.

THERMAL POLLUTION CONTROL TECHNOLOGY

Contrary to the prevailing situation with SO_2 or heavy metal emissions, adequate and reliable technology exists today for the control of thermal pollution. The only barrier to the wide adoption of cooling towers or cooling ponds is the reluctance of the electric utility industry in the face of their relatively small extra contribution to the cost of power.

Cooling Ponds

A cooling pond is a pond or lake especially constructed to provide a source of cooling water and a closed sink to which to return it after use. The pond must be sufficiently large so that evaporation and other heat transfer to the air from its surface can keep the average water temperature low enough to provide a continuous source of cooling water. The pond outlet is usually located as far as possible from the hot water return to maximize the time allowed for cooling.

Unfortunately, from 1 to 2 acres of cooling pond are required per MWe of plant capacity, depending upon plant type and efficiency, average ambient temperature, local winds, and lake depth. Since 1000 to 2000 acres of land are required for the cooling pond for a modern 1000 MWe plant, their use is restricted to regions where land is both cheap and available. The cost of cooling pond installation has been variously estimated at from $2.50 to $12/KW capacity above the cost of once-through cooling (Lof and Ward, 1970; Woodson, 1969; Eicher, 1969).

Various schemes have been proposed for increasing the productivity of cooling ponds, including use of ponds for recreation (Eicher, 1969) and for raising fish at accelerated rates (Mathur and Stewart, 1970). Questions remain about the possibility of chemical contamination of such fish, or even about possible high levels of radioactivity in fish raised in cooling water, which is cycled repeatedly through a nuclear power plant. Overstimulation of aquaculture may also cause problems with the disposal of the waste products from high fish concentrations (Mathur and Stewart, 1970).

Cooling Towers

A cooling tower is a large device which allows for the efficient contact of ambient air with heated water, resulting in the rapid cooling of the water without or before returning it to its source. When a cooling tower is used, it is possible to use the same water for power plant cooling again and again, by adding a small fraction of fresh water to make up for evaporation losses, "blow down," and "drift." Water withdrawal rates can be reduced to 2–4% or less of those required for once-through cooling (Lof and Ward, 1970).

The details of cooling tower design have been reviewed recently (Woodson, 1971) and they will not be discussed here. However, we do need to understand a few broad terms to complete our discussion. Cooling towers are classified by design as either "wet" or "dry" and as being "natural" or "mechanical" draft. In a wet tower the heated water is broken into a fine spray which falls through moving ambient air. The water drops cool by evaporation of a few percent of their weight, reaching a lower temperature which is theoretically bounded by the ambient temperature and humidity. The lower limit of temperature is the "wet-bulb" temperature. The actual temperature reached by the water is higher than the wet bulb temperature by an amount which depends on the tower design details.

Dry cooling towers function much like large auto radiators in which the heated water passes through closed coils over which ambient air moves. Heat is transferred through the coils by conduction and into the air by convection. No water is lost by evaporation and the lowest temperature attainable by the water is the ordinary "dry bulb" temperature, which is higher than the wet bulb temperature and is the temperature recorded by an ordinary thermometer.

In mechanical draft towers the flow of air is driven by large fans, whereas the air flow in a natural draft tower is driven by the chimney effect of bouyancy differences between cool and heated air. Mechanical draft towers are large boxy affairs while natural draft towers are tall and usually hyperbolic in cross section.

Cooling towers for steam electric plants are quite large. A mechanical draft tower for a 1000 MWe plant may be 70 feet wide, 70 feet tall, and 300–400 feet long. A hyperbolic tower for the same plant would be upwards of 400 feet tall and 400 feet in diameter.

The American electric utility industry has used mechanical draft wet towers for many years, and has used natural draft, hyperbolic wet towers since 1962. Eighteen natural draft towers were in service in mid-1970 and 32-40 of them were completed by 1972-73 (Stockham, 1971). Only two very small mechanical draft dry towers have been in operation at the Neil Simpson power plant of the Black Hills Power and Light Company at Wyodak, Wyoming; one on a 3 MWe unit since the early sixties, and a second one on a 20 MWe unit since 1969. No natural draft dry towers have been installed on commercial power generating plants in the United States. The report by Rossie and Cecil (1970) is a comprehensive review of dry cooling tower technology and of European experience with dry towers of up to 200 MWe capacity.

Environmental Costs of Cooling Tower Use

The major benefit to be gained from the use of a cooling tower is the elimination of, or very substantial reduction of, heated water discharge to rivers, lakes, or estuaries; but this benefit is not gained without some environmental costs.

Dry towers are nearly pollution free, with the exception of the discharge of heated air. This ultimate discharge of heat to the atmosphere cannot be ignored in the long run since it is the inevitable result of thermal electric power generation. In fact, all of man's use of energy leads eventually to heat release according to the second law of thermodynamics. The large scale effects of man's heat release on the earth's climate has been considered in the short term by the Committee for Environmental Information (1970) and the Study of Critical Environmental Problems report (1970) and for the longer term by Porter and others (1971). Their calculations indicate that global heating will eventually limit our use of even "free" fusion power. However, the capacity of the atmosphere to accept thermal discharge without major harm is now thought to be much greater than that of inland waterways or estuaries.

As we shall see in the next section, the use of a dry tower may tend to reduce the overall efficiency of a power plant somewhat on exceptionally hot days so that more fuel must be burned to produce the same electric power output. The additional air pollution and resource depletion due to the extra fuel consumed should be considered an environmental cost of dry tower service. Nevertheless, dry cooling towers are felt to be the best available solution to waste heat disposal by many environmentalists.

Three principal environmental problems can be cited as costs of wet cooling towers. These are water loss by evaporation and drift, the discharge of treatment chemicals in the "blow down" stream, and the effects of the moist heated air plume. Indirectly, wet tower use may cause a slight drop in plant efficiency and a resulting increase in air pollution and fuel use. Under the conditions of coincident hot weather and peak power demand, estimates of loss of plant capacity range from 0.5 to 1.0% for wet towers and from 3 to 14% for dry towers (Rossie and Cecil, 1970; Rossie, 1971; Romm, 1971). These losses occur because energy is required to operate the towers' fans (if any) and pumps, and because at high plant loading and ambient temperature the tower may not be able to cool the circulating water as much as is desirable.

Water by evaporation in a wet cooling tower is about 1% of throughput for each 10°F drop in water temperature in the tower (Thompson, 1968). In addition, somewhat less than 0.2% of throughput is lost as drift, or liquid water drops carried out of the tower in the exit air stream. For a 1000 MWe plant operatiing at a 20°F rise through the condenser and 20°F drop in the cooling tower, evaporation loss will be about 21 cfs and drift loss less than 2 cfs. Total flow through the cooling tower will be about 1050 cfs.

To prevent corrosion and fouling of the condenser and the growth of algae, fungus, and bacteria in the cooling tower, it is necessary to add various treatment chemicals to the cooling water. Among the many chemicals which may be added (although not all in the same installation) are chromates, phosphates, sodium silicate, zinc, nitrates, borates, film forming polymers, natural organic dispersants, chelating agents, organotin compounds, organosulfur compounds, chlorine, hypochlorites, chlorinated phenols, quaternary ammonium compounds and complex amines, and organo-thiocycantes (Silverstein and Curtiss, 1971; FWQA, 1970). To prevent buildup of those compounds as well as dissolved solids left behind by evaporation, it is necessary to continuously or intermittently discharge a stream of cooling water, usually back to the source. This stream, the so-called "blow down," is usually about 0.3% of tower throughout per 10°F of cooling (Thompson, 1968). For our 1000 MWe plant with a 20°F rise the blow down will be on the order of 6 cfs. The blow down can be withdrawn from the cooling tower discharge and returned to the source at nearly its original temperature.

To replenish the evaporative drift, and blow down losses, then, requires a flow of "make-up" water from the source of about 29 cfs, or about 3% of the requirement of 1000 cfs for "once-through" cooling.

The third source of concern with wet cooling towers and the environment is the discharge of the warm, moist plume. During normal operation the water vapor forms a visible plume which may persist from several yards up to several miles from the tower, depending upon weather conditions. While a number of authors have speculated on the ability of the plume to enhance local precipitation or cause other local weather anomalies, a recent survey of the situation at the 1800 MWe Keystone Station in Pennsylvania could detect no effect on weather locally or several miles away (Stockham, 1971). During that study it was noted that SO_2 from the station smokestacks could co-mingle with the tower plume to form sulfuric acid droplets in the plume. The possible effect of this phenomenon has not been evaluated.

Finally, the question of the aesthetics of cooling towers has been raised. They are said to be large and offensive to the public eye. However, this point is invariably raised by power industry spokesmen, not by environmentalists. This point may be being raised in desperation or perhaps as a throwback to the days when the public's chief concern with pollution was often only objection to visible stack plumes.

Economics of Cooling Towers

Many articles and reports have dealt with the cost of cooling tower construction and operation. While the costs of a tower can vary widely with location, the published figures reflect a larger diversity of opinion on the subject than this factor alone can account for. Among the factors which influence a cost estimate for a given location are:

1) Type of tower.
2) Whether the tower is added to an existing plant design which has been optimized for once through cooling, or whether the entire design is optimized with a cooling tower. The former practice can unfairly inflate the calculated cost of a tower, particularly for dry towers which must be severely over-designed to accommodate steam turbines of conventional design.
3) Whether the quoted cost is given credit for making the installation of a once-through system unnecessary.
4) Whether the tower is charged a penalty for loss of peak generating capacity or for extra fuel to maintain capacity required. The assumed marginal cost of peaking capacity can be of critical importance.
5) Whether the use of a tower is given credit for reducing the cost of cooling water, if any, or is given any dollar credit for eliminating thermal discharge.
6) Use of a cooling tower may introduce power plant siting flexibility by eliminating the need for a large source of cooling water. This independence may permit savings in fuel or power transporation costs.

Typical capital cost data for cooling towers of various configurations have been compiled from a number of sources and are summarized in Table 1. The data are broken down into four categories to show how various assumptions can lead to widely different cost estimates for a given tower configuration. Within the categories the cost figures given by various authors tend to agree, with the exception of Woodson's estimates (1969) for dry cooling towers which appear to be on the high side.

For discussions of cooling tower installation at proposed new power plants, it is category III which is most relevant, since all power plants must be equipped with at least once-through cooling systems.

Some typical figures for the increase in the cost of power generation are summarized in Table 2. It is especially difficult to compare these kinds of data from the literature due to broad variations in the assumed cost of operation, interest rates, etc. For comparison purposes, total power generating costs usually lie between 3 to 7 mills/kwh. Many estimates of the additional power cost due to cooling tower operation to the average residential consumer lie in the range 1/2 to 5% for wet towers and 2 to 7% for dry towers. It should be kept in mind that generation cost usually accounts for less than one-half of ultimate residential-consumer power cost.

Table 1 Typical Cooling Tower Investment Costs, $/KW Plant Capacity[a]

Plant Type	Fossil Fuel				Nuclear Fuel			
Cooling Tower	Wet		Dry		Wet		Dry	
Type	Mechanical	Natural	Mechanical	Natural	Mechanical	Natural	Mechanical	Natural
Cost Basis								
I	9	12	22–24	19–40	14	19	27–45	65
II	5–9	6–12	15–22	17–39	8–13	9–20	20–33	22–63
III	2	5	16–19	21–33	4	9	27	55
IV	2–5	4–7	13–17	18–33	3–6	6–8	21–23	25–53

Data Sources: Jimeson and Adkins, 1971; Woodson, 1969; Rossie and Cecil; FWQA, 1970; Dickey and Cates, 1970.

Cost Bases Descriptions

I Total cost including extra peak power capacity cost
II Total cost without extra peak power capacity cost
III Additional cost above once-through cooling including extra peak power capacity cost
IV Additional cost above once-through cooling without extra peak power capacity cost

[a]Costs are typical for steam electric generating plants of the 500-2000 MWe class.

Table 2 Typical Increase in the Cost of Power Generation Due to Cooling Tower Use mil/KWH (Additional Costs Above Cost of Once Through Cooling)

Plant Type	Fossil Fuel				Nuclear Fuel			
Cooling Tower	Wet		Dry		Wet		Dry	
Type	Mechanical	Natural	Mechanical	Natural	Mechanical	Natural	Mechanical	Natural
Costs	.08–.20	.14–.22	.46–.80	.43–.98	.09–.2	.2	.8–.9	.8–1.4

Data Sources: Woodson, 1969; FWQA, 1970; Oleson and Boyle, 1971.

BENEFICIAL USES OF WASTE HEAT FROM ELECTRIC POWER PLANTS

Frequently it is suggested that we should regard the vast quantities of waste heat from electric power generation not as a source of pollution, but as a national resource to be employed in accomplishing useful tasks. (Mathur and Stewart, 1970; Miller et al., 1971). Among the beneficial uses which have been suggested are heating (and cooling) of large cities (Miller et al., 1971) or of new towns (Beall, 1970); desalination of water and industrial process heating (*Chem. Eng.*, 1971); heating green houses, vegetable farms, or fish ponds (Mathur and Stewart, 1970; Beall, 1970; Yee, 1970); heating of rivers which normally freeze to lengthen navigational seasons (Biggs, 1968); defogging of airports or deicing of runways and highways; and puncturing of atmospheric thermal inversion layers (Smith, 1971).

Some of these schemes are practical, in fact, urban generating stations have provided heat to nearby buildings in New York, St. Louis, and other cities for years. Since modern stations are so large and since new urban construction is relatively scattered, it is not so convenient to provide this service from new power plant construction. Beall (1970) has suggested, therefore, that so-called "new towns" be designed and built near generating stations of appropriate size.

There is a practical problem associated with a number of the ideas, which is that all the waste heat from power plants as currently designed is available at temperatures between 85° and 105°F. It is a fundamental fact of physics that heat only flows from high to low temperatures, so that waste heat cannot be used to raise the temperature of anything above its own temperature by conduction. As a result it has been suggested that expanding steam be discharged from the turbine at higher pressures and temperatures, say at 150° to 180°F. A trade-off is experienced in which the electricity generating efficiency drops while the efficiency of waste heat utilization is increased. The overall "total energy" system may have a higher thermodynamic and economic efficiency than would the power plant alone.

Other problems associated with these ideas include the fact that the power plant always produces heat whether the "beneficial use" needs it or not, making alternate thermal discharge facilities necessary. A corollary to the problem of heat availability is that tremendous quantities of warm water must be handled. For example, to transport the 1000 cfs of heated water from our 1000 MWe plant at a reasonable speed of 10 feet per second would require a pipe over 11 feet in diameter. Such large ducting and the necessary pumps and fixtures are quite expensive for transport of water over all but the shortest distances.

Whereas it is to society's advantage to make use of waste heat where economically and environmentally sound, it is important to realize that all such heat eventually is released to the larger environment and that the global limit on heat release is not thereby circumvented.

CONCLUSIONS

Discharge of waste heat from steam electric generating plants to waterways is a significant and rapidly growing problem. Fortunately, either cooling towers or cooling ponds can be used to transfer waste heat to the atmosphere, where less environmental degradation is likely to occur. Dry cooling towers have less environmental impact than wet towers since the evaporation loss from wet towers is eliminated. Eventually, a limit will be reached on overall thermal discharge by any technique, due to global heating.

Wet cooling towers for a modern 1000 MWe generating plant cost from $2 to $9 million, while dry towers for the same plant may cost $16 to $55 million. No large dry towers have been used in the United States. Use of cooling towers may cause an increase of 1/2% to 7% in the average residential electric bill.

A number of suggestions have been presented for beneficial use of waste heat, but most are still in the planning stage. The ultimate problem of a limit to global thermal discharge cannot be avoided by using waste heat "beneficially."

REFERENCES

Beall, S. E., Jr. 1970. *Reducing the environmental impact of population growth by the use of waste heat.* Oak Ridge Laboratory Report. Dec.

Biggs, J. G., 1968. *Nuclear station waste heat can extend the St. Lawrence Seaway season.* Atomic Energy of Canada, Ltd., Report No. AECL-3061, Feb.

Bregman, J. I., 1969. Quoted by P. A. Krenkel and F. L. Parker in Chapter 2 of *Biological aspects of thermal pollution,* eds. P. A. Krenkel and F. L. Parker. Nashville: Vanderbilt Univ. Press.

Chemical Engineering, 1971. Study finds energy center feasible for Puerto Rico, p. 58. Aug. 9.

Committee for Environmental Information, 1970. The space available. *Environment* 12 (Mar.).

Dickey, J. B., Jr., and R. E. Cates, 1970. *Managing waste heat with the water cooling tower.* Kansas City, Mo.: The Marley Co.

Eicher, G. J., 1969. Cooling lakes can be a pleasant solution. *Electric World,* Apr. 14:90.

Eisenbud, M., and G. Gleason, eds., 1969. *Electric power and thermal discharge,* New York: Gordon and Breach.

Federal Power Commission, 1969. *Problems in disposal of waste heat from steam-electric plants.*

———, 1971. *Steam electric plant construction cost and annual production expenses.* Thirty-second Annual Supplement 1969.

Federal Water Pollution Control Administration, 1968a. *Report on water quality criteria.* Apr.

———, 1968b. *Industrial waste guide on thermal pollution control.* Sep.

Federal Water Quality Administration, U.S. Dept. of the Interior, 1970. Feasibility of alternate means of cooling for thermal power plants near Lake Michigan. Aug.

Jaske, R. T., and C. J. Touhill, 1970. An independent view of the use of thermal power station cooling water to supplement inter-regional water supply.

Jimeson, R. M., and G. G. Adkins, 1971. Waste heat disposal in power plants. *Chem. Eng. Progr.* 67, no. 7: 64.

Krenkel, P. A., and F. L. Parker, eds., 1969. *Biological aspects of thermal pollution.* Nashville: Vanderbilt Univ. Press.

Lof, G. O. G., and J. C. Ward, 1970. Economics of thermal pollution control. *J. Water Pollution Control Fed.* 42: 2102.

Mathur, S. P., and R. Stewart, 1970. Proceedings of the Conference on Beneficial Uses of Thermal Discharges, Sept. 17-18, 1970. New York State Department of Environmental Conservation.

Miller, A. J., et al., 1971. *Use of steam electric power plants to provide thermal energy to urban areas.* Oak Ridge National Laboratory Report No. ORNL-HUD-14, Jan.

Oleson, K. A., and R. R. Boyle, 1971. How to cool steam electric power plants. *Chem. Eng. Progr.* 67, no. 7: 70.

Porter, W. A., et al., 1971. Global temperature effects of the use of fusion energy and the fusion torch. *IEEE Nuclear Transactions* Feb., p. 31.

Romm, J., 1971. The cost of cooling towers for Bell Station. Report from Cornell University Water Resources and Marine Sciences Center, Ithaca, N.Y. Mar. 1969.

Rossie, J. P., 1971. Dry type cooling systems. *Chem. Eng. Progr.* 67, no. 7: 58.

Rossie, J. P., and E. A. Cecil, 1970. Research on dry-type cooling towers for thermal electric generation: part I. Contract No. 14-12-823 for the Water Quality Office, Environmental Protection Agency, U.S. Gov't. Document EP210:16130EES 11/70, Nov. 1970.

Silverstein, R. M., and S. D. Curtiss, 1971. Cooling water. *Chem. Eng.* Aug. 9, p. 84.

Smith, G., 1971. G. E. plans stadium-like tower, *New York Times*, Apr. 7, 1971.

Stockham, J., 1971. *Cooling tower study.* I.I.T. Research Institute, Report No. C6187-3 Jan. Prepared for Environmental Protection Agency, Air Pollution Control Office.

Study of Critical Environmental Problems, 1970. *Man's impact on the global environment.* Cambridge, Mass.: MIT Press.

Thompson, A. R., 1968. Cooling towers. *Chem. Eng.* Oct. 14, p. 100.

U.S. Dept of the Interior, 1961. Surface water supply of the United States 1960. Part I-A. Paper No. 1701.

Woodson, R. D., 1969. Cooling towers for large steam-electric generating units. In Eisenbud and Gleason, q.v.

———, 1971. Cooling towers, *Sci. Am.*, 224: 70.

Yee, W. C., 1970. Food values from heated wastes—An overview. Reprint of paper for the Proceedings of the 32nd Annual Conference of the Chemurgie Council, Washington, D.C. Oct. 22–23, 1970.

ROBERT FRANK

Biologic Effects of Air Pollution

Air Pollution arises chiefly from the combustion of fossil fuels. Its composition will vary with the type of fuel used, the method and efficiency of combustion, and whatever techniques of emission control are applied. Nevertheless, there are five classes of pollutants released whenever fossil fuels are consumed. They are: oxides of sulfur, oxides of nitrogen, particles (sometimes referred to as aerosols or fly ash), hydrocarbons, and carbon monoxide (CO).

In many parts of the country electric power plants are major sources of pollution. It has been estimated that power plants in the United States emit by weight about 50% of all oxides of sulfur, 30% of all oxides of nitrogen, 20% of all suspended particles, and 0.8% of all carbon monoxide. These emissions, either alone or in combination, are capable of damaging health, reducing safety in the performance of certain tasks, and offending our sense of well-being. The adverse effects of pollutants on man and his environment are referred to legislatively and legally as air quality criteria.

The effects of urban air pollutants on health, except for those caused by a few specific elements such as CO, lead, and beryllium (NAPCA publications; EPA, 1971; NAS and NAE, 1969; Higgins and McCarroll, 1970), are neither discrete nor readily quantified. Urban pollution probably acts in combination with other forms of stress rather than independently. Its principle effect may be to worsen rather than initiate disease (Lawther, 1966). The other forms of stress that enter into this complex interplay include socio-economic factors (crowding, poor hygiene, malnutrition), weather (especially coldness, dampness, and perhaps sudden, extreme changes in temperature), and most importantly, tobacco smoking. Smoking, a personal form of air pollution, is considered the greatest cause of chronic respiratory disease and cancer. Unfortunately, it has frequently confounded efforts to assess the role of pollution on health because of its overwhelming toxicity. In fact, studies of the impact of air pollution on children have been undertaken in part for the purpose of avoiding this complication, as well as the complication of occupational exposure.

Evidence of the biologic effects of air pollution, exclusive of damage to vegetation, derives from three main avenues of investigation:[1]

1. TOXICOLOGIC STUDIES ON ANIMALS

While these studies have been valuable for clarifying the mechanisms of biologic response and for describing dose-response relations, they have proved of only limited use in providing criteria for the establishment of ambient and emission standards. The overwhelming majority have focused on acute, essentially reversible effects (if the animal survived) rather than on chronic exposures which might lead to cumulative and persistent damage to the lungs or other organs. Perhaps their principal shortcoming has been a reliance on exposures too simple in composition and/or unrealistically high in concentration. For example, the levels of sodium dioxide (SO_2) in urban atmospheres associated with increasing morbidity and mortality have virtually no effect in laboratory investigations when the gas is administered in otherwise clean air. Since there is no certainty about which individual pollutant or combinations of pollutants may be critical in urban pollution,[2] the toxicologist has been understandably reluctant to generate a complex mixture of gases and particles, particularly for use in chronic studies, which may ultimately prove irrelevant or benign. The studies are also beset with the perennial uncertainty over how readily their results can be extrapolated to man.

2. EXPOSURE OF HUMAN SUBJECTS TO CONTROLLED DOSES OF POLLUTANTS

Human exposures have usually lasted up to several hours. They share some of the advantages and disadvantages of the studies on animals. While such studies have yielded information about the mechanisms of

[1]An excellent critique of these types of investigation is to be found in National Institute of Environmental Health Sciences, 1970.

[2]There are exceptions to this statement, as for example, the levels of CO found in the vicinity of heavy automotive traffic, and those of fly ash and oxides of sulfur immediately downwind to large industrial sources. But these circumstances are analogous to the acute severe episodes of air pollution rather than the more chronic, low-grade forms to which larger numbers of persons are exposed.

response and the probable importance of individual pollutants, they have frequently had to resort to unrealistically high concentrations of one, or at most, a combination of two pollutants to evoke a response. Recent experiments which combined the pollutant with an additional stress such as exercise have elicited responses at more realistic concentrations (Bates et al.). Although potentially a valuable source of criteria, the studies have yielded disappointingly little information in the past. Several factors contribute to this poor record. Perhaps the most important is the ethical dilemma posed by the use of human subjects, particularly those having pulmonary or cardiovascular disease. Although most of the epidemiologic evidence for the adverse effects of air pollution on health have been obtained on such patients, there is monumental unease over subjecting sick persons to pollutants experimentally, even under controlled laboratory circumstances. Until recently, this form of applied research on man and laboratory animals attracted relatively few competent investigators. Fortunately, there has been a recent influx of capable investigators into the field as concern about the environment has intensified, and the financial support for research has grown.

3. EPIDEMIOLOGIC SURVEYS AND COMMUNITY HEALTH

Acute, severe episodes of air pollution in combination with specific patterns of weather have been associated with increased rates of morbidity and mortality, particularly among the aged and sick. Again, however, the possible role of chronic, low grade pollution on health remains uncertain. While community studies have provided the bulk of our criteria, they have yielded only fragmentary information on the specific pollutants responsible, as well as their concentrations. One great difficulty has been to estimate the dose of a specific pollutant to which the individual is exposed. There may be considerable gradients in concentration between the indoors and outdoors, within a small outdoor area (vertical and horizontal gradients are not uncommon), and from one neighborhood to the next. The complexity of the problem increases enormously when one attempts to estimate doses for extended periods of time, particularly for a population that is becoming increasingly mobile.

Vulnerability to air pollution varies widely among different segments of the population. The very young and the elderly appear to be especially vulnerable. DuBos and co-workers (1968) have shown in experimental animals that the effects of pharmacologic agents on the fetus at or near term and on the newborn, are exaggerated in magnitude and duration. Whether man at the same stages of development is equally vulnerable to chemical stress, including that of air pollution, is unknown at present. The difficulty of obtaining such vital information directly will be mammoth, but the problem does deserve assiduous investigation both epidemiologically and experimentally. The problem might be more amenable to solution if we increased our capacity to accumulate and analyze medical, health, and air sampling information on a national scale, and were prepared to cope with mobile populations; that is, some equivalent of large data banks would be most helpful. Pregnancy and childhood are two additional periods of life when susceptibility to pollutants may increase. The association between air pollution and lower respiratory infection from infancy into adolescence is quite impressive (Douglas and Waller, 1966; Holland et al., 1969). Whether this vulnerability is innate or may be related, at least in children, to increased activity outdoors is not known.[3]

The most susceptible persons are likely to be those already ill, particularly with chronic respiratory and/or cardiovascular disease. The dramatic increases in morbidity and mortality rates associated with severe episodes of air pollution have traditionally centered on these patients. But there is evidence that less marked fluctuations in urban air pollution may also aggravate their illness. A recent survey in Chicago showed that elderly patients[4] with moderate-to-advanced chronic bronchitis suffered an increase in symptoms during and immediately after any increase in local pollution for which SO_2 and particulates were used as the indices (Carnow, 1970; et al., 1969). Perhaps implicit in this observation is the notion that we shall probably continue to unveil greater evidence of the harmful effects of air pollution as our methods of study become more sensitive and sophisticated, and as they are applied more extensively. The question that may confront society as our knowledge is expanded is not, "What is a safe level of pollution?" but rather, "What is a tolerable or acceptable level of sickness from pollution?"

A considerable variety of biological effects are imputed to air pollutants.[5] A partial list follows:

[3]The increased ventilatory rate and obligatory mouth-breathing characteristic of exercise and heavy work are equated with increased exposure of the lungs. The increased exposure occurs in part because the upper airways become less efficient gas scrubbers as ventilatory rates increase.

[4]The elderly and chronically ill segments of our society are likely to grow in size to the extent that improved medical and community services of the future help increase the average span of life. These groups are among the most vulnerable to the toxic effects of pollution.

[5]Concerning the effects on vegetation see Chapters 3 and 4 of this volume, and the publications of NAPCA; EPA (1971); NAS and NAE (1969); and *Power Plant Siting and Air Quality*.

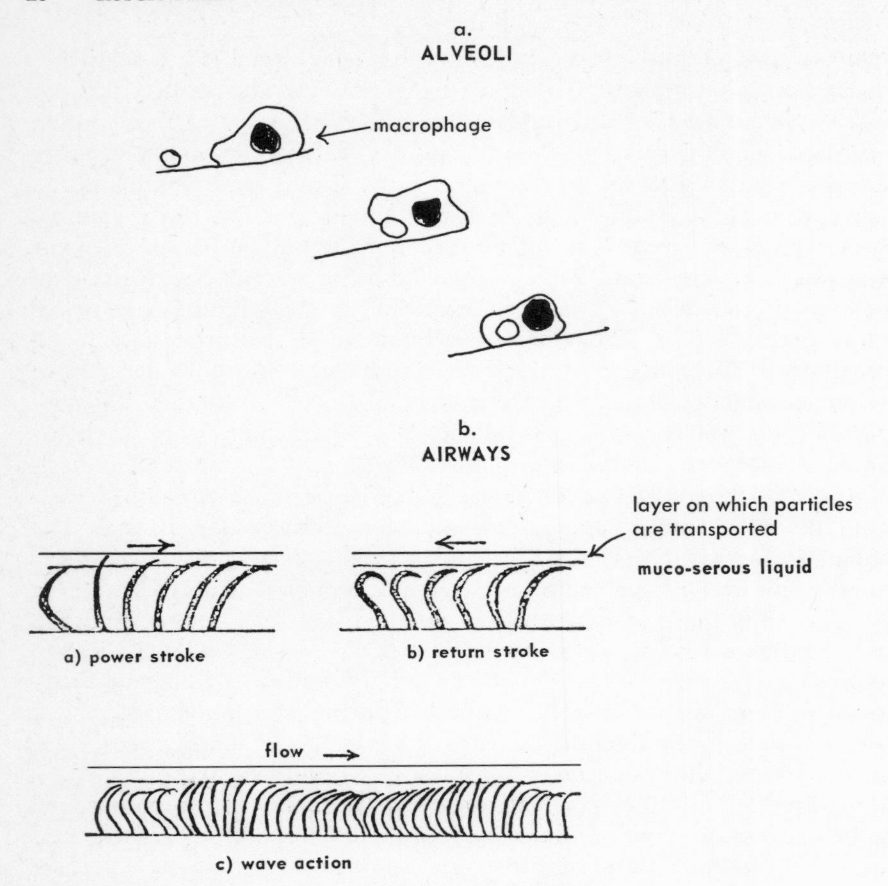

Figure 1 Housecleaning ability of the healthy lung
The healthy lung is virtually sterile, accomplishing its housecleaning with two linked elements. One is the alveolar macrophage, a mobile scavenger cell. Macrophages engulf particles, animate and inanimate, that land on alveolar surfaces. These cells have enzymes capable of killing and digesting microorganisms. The second element is the muco-ciliary system. The cilia are hair-like projections lining the airways of the mouth, nose, pharynx, and tracheobronchial tree and lie immersed in muco-serous liquid. They beat about a 1,000 times each minute, their whip-like beat propelling the liquid on which foreign material is supported to the pharynx where it is swallowed or expectorated. (See Denton et al., 1965.)

1. Reduced Resistance to Infection of the Lung. The occurrence and severity of respiratory infections in laboratory animals is increased, and the capacity of the animal to inactivate or kill inhaled organisms is impaired following exposure to oxides of nitrogen (Ehrlich and Henry,

1968; Henry et al., 1970) or ozone (Coffin and Blommer, 1965). The basis for these effects may be severalfold. Ozone, in particular, damages the alveolar macrophage, a cell (Fig. 1a) that functions to engulf, inactivate, and dissolve invasive organisms (Coffin et al., 1968). Both gases may also damage the epithelial lining of the airways and thereby alter mucociliary function (Fig. 1b): the latter is vital in cleansing dust and bacteria from the lung. It is notable that the destruction of inhaled bacteria is also impeded by malnutrition, acute alcoholism, and smoking which, of course, co-exist in the community with pollution. These different forms of stress may be expected to reinforce each other's toxic potential. Paradoxically, while the concentrations of photochemical oxidants encountered in the urban and occupational environments can be shown to reduce the body's resistance to infection in laboratory animals, there is virtually no evidence for such an effect in human populations. An equal but opposite paradox applies to SO_2 and particulates; that is, the association between the latter pollutants and respiratory infection in community studies is strong, whereas laboratory confirmation, based on realistic dosages of these agents, has been wanting.

2. *Aging and Chronic Disease of the Lung.* Collagen and elastin, the two most abundant structural proteins of the lung, contribute importantly to its elastic behavior and structural integrity. Collagen, and probably elastin as well are denatured by nitrogen dioxide (NO_2) and ozone (Buell, 1965). As a consequence, the molecular configuration is thought to have important implications for the rate at which the lung ages structurally, and may also provide a basis for degenerative processes that culminate in diseases like emphysema. There is evidence, for example, that the lifetime exposure of laboratory animals to concentrations of NO_2 as low as 2 ppm may cause irreversible damage to the lungs, a damage bearing resemblance to human emphysema (Freeman et al., 1968). Whether man is equally sensitive to these gases is unknown. To isolate such effects epidemiologically is probably beyond our present abilities.

3. *Carcinoma of the Lung.* While death rates from carcinoma of the lungs are higher in cities than in rural areas, the role of air pollution in this difference is uncertain (Anderson, 1967). Aromatic hydrocarbons such as 3,4 benzpyrene, which induce carcinoma experimentally, are present in urban air (and cigarette smoke). Two important sources for the carcinogens are auto exhaust and coal-burning plants. These hydrocarbons, and perhaps other elements in urban air, probably play an adjunctive role, rather than a primary one, in the complex process leading to carcinoma.

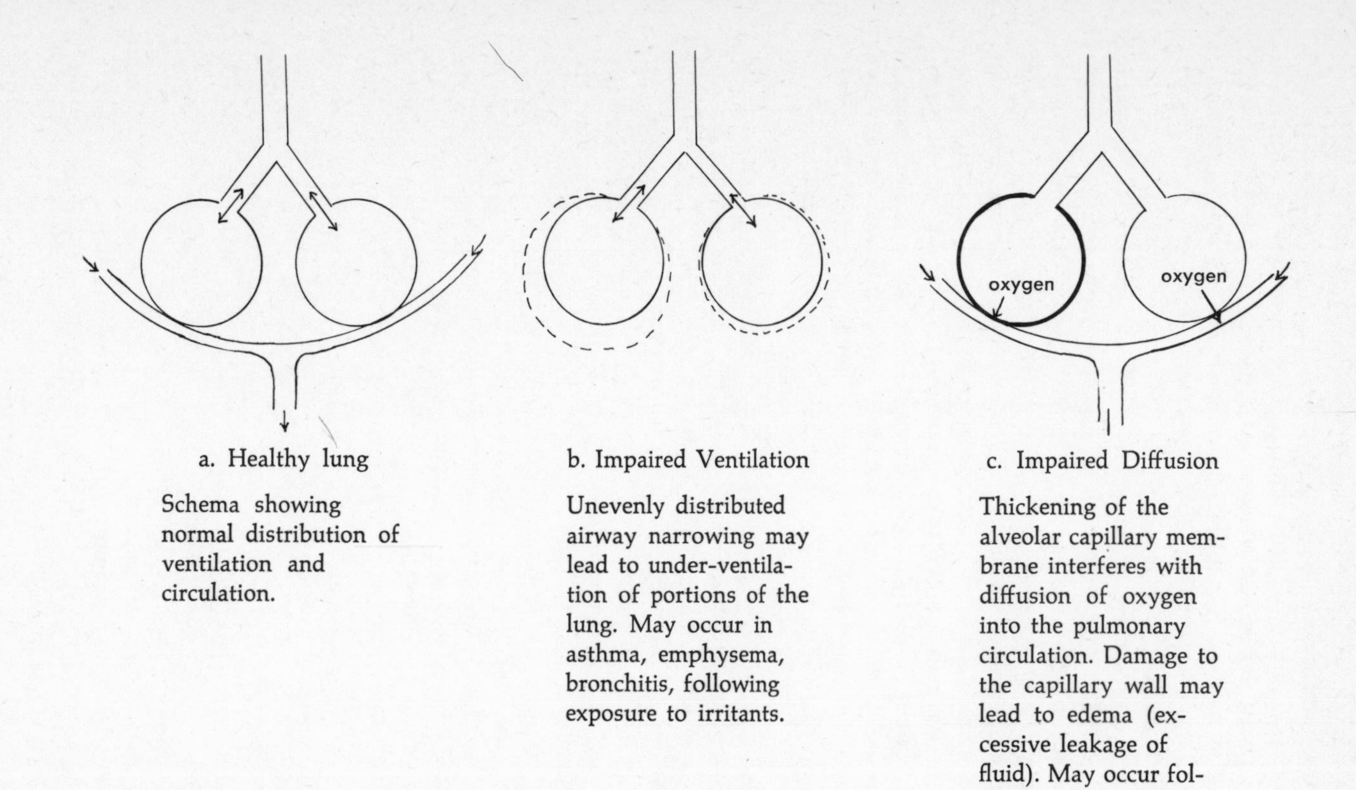

a. Healthy lung

Schema showing normal distribution of ventilation and circulation.

b. Impaired Ventilation

Unevenly distributed airway narrowing may lead to under-ventilation of portions of the lung. May occur in asthma, emphysema, bronchitis, following exposure to irritants.

c. Impaired Diffusion

Thickening of the alveolar capillary membrane interferes with diffusion of oxygen into the pulmonary circulation. Damage to the capillary wall may lead to edema (excessive leakage of fluid). May occur following exposure to O_3, NO_2.

Figure 2 Impaired ventilation and diffusion in the lung

Laboratory Experiment

SO_2 (gas phase) + Aerosol (physiologically inert) ⟶ Irritant Aerosol

Natural Atmosphere

SO_2 (gas phase) $\xrightarrow{\text{oxidized}}$ H_2SO_4, sulfates incorporated in submicronic particles

Figure 3 Production of irritant aerosols
SO_2 alone is removed by the upper airways. The irritant aerosol may be deposited in the tracheo-bronchial tree of alveoli, depending on its aerodynamic properties. In the natural atmosphere, the reaction is catalyzed by heavy metals (lead, mercury, etc.), photochemical pollutants, sunlight, and dusts. In the laboratory, the conversion of the aerosol to a droplet by raising relative humidity increases the effect.

4. *Altered Respiratory Mechanics.* The lung is designed structurally to distribute inspired air over a diffusing surface that is about 60–80 square meters in area, and comprises about 300 million alveoli.[6] The conducting airways, starting at the trachea, may undergo more than 20 dichotomous divisions or branchings before terminating at the alveoli. Despite such structural complexity, the inspired air is distributed evenly at little cost in energy. This remarkable performance requires that the frictional resistance to airflow remain slight and be approximately equal among the many parallel airways. Both requisites are met in the normal lung (Fig. 2a). Inhaled irritants may provoke either localized or widespread narrowing of the airways and so cause an increase in the work of breathing. If, as frequently happens, the constriction is non-uniform, the air that is inhaled may be unevenly distributed throughout the lung (Fig. 2b). When these defects become severe, shortness of breath and inadequate exchanges of oxygen and carbon dioxide across the lung may become disabling. Sodium dioxide, nitrogen dioxide, ozone, certain hydrocarbons, and particulates are among the irritants capable of producing these effects. The magnitude of the response will depend on the concentration of the irritant in inspired air, the mode of breathing—breathing by mouth at an increased rate, as in exercise, is equated with an increased exposure of the lower airways (Frank et al., 1969)—and the reactivity of the tracheo-bronchial tree (Fig. 3).

[6]The alveolus is the functional unit of the lung, the site of exchange of oxygen and carbon dioxide. It is extremely thin walled (less than 1 micron thick), and about 150 microns in diameter.

Patients with asthma or tracheo-bronchitis may react excessively and experience bronchospasm amid levels of pollutants that have no appreciable effect on healthy individuals. The most conspicuous laboratory evidence for the synergistic effects of mixtures of pollutants, namely, SO_2 and sodium chloride aerosol, has been obtained with these measurements of airway mechanics (NAPCA, 1969; Amdur, 1961).

There is uncertainty over the relation between short-term airway narrowing and either the capacity of the lung to resist infection or the occurrence and progression of chronic irreversible disease. There is little doubt however, that the superimposition of new or aggravated mechanical defects on a lung that is already diseased can be seriously disabling.

5. Interference with Oxygenation of Tissue. There are several ways whereby air pollutants may cause a reduction in the supply of oxygen to the tissues of the body. Mention has already been made of the impairment in the distribution of air in the lung that may attend airway narrowing. Another way is by damage to the alveolar-capillary membrane. The latter separates the air in the alveoli from the neighboring circulation. It is the path across which oxygen and carbon dioxide move to enter and leave the body. The extreme thinness of this membrane enables the lung to serve as an efficient diffusing surface. The transfer of oxygen is impaired whenever this membrane is thickened by inflammation, edema (leakage of fluid from the blood vessels), or scarring (Fig. 2c). Low concentrations of ozone may damage the alveolar-capillary wall and produce edema. Ozone, perhaps the most damaging component of photochemical smog, has been shown to diminish (reversibly) the diffusing capacity of the lung in healthy volunteers (Bates et al., 1972; Young et al., 1964).

A third source of interference with the supply of oxygen to tissues is carbon monoxide, an asphyxiant. Carbon monoxide (CO) has an affinity for hemoglobin over 200 times greater than that of oxygen (NAPCA, 1969, 1970). Therefore, it displaces oxygen from hemoglobin, competitively. Moreover, the oxygen that does combine with the hemoglobin is less readily released to the peripheral tissues when CO is also present as carboxy-hemoglobin (CO · Hb). These two factors, by decreasing the volume of oxygen at the tissue level, may seriously interfere with the function of peripheral tissues. The two organs most sensitive to oxygen deprivation are the brain, especially at the cortical level, and the heart. In healthy volunteers, CO · Hb levels as low as 2–5% have reportedly been associated with changes in visual and time perception, and in certain psychomotor tests (NAPCA, 1969, 1970). Such changes might interfere with performances requiring a high degree of skill and attention. The population at greatest risk includes individuals who are already hypoxic from a variety of diseases or circumstances: anemia, chronic pulmonary or cardiac disease, and high altitude, or those who have increased oxygen needs as in pregnancy, fever, and hyperthyroidism. The association between cigarette smoking and increased mortality from coronary vascular disease may partly reflect the increase in CO · Hb that is typical of smokers (Goldsmith et al., 1969). The fetus may be particularly vulnerable to asphyxia from CO; thus, newborn babies of mothers who smoke appear to be smaller than those of non-smoking mothers, a difference that may be related to the CO · Hb shared by both the maternal and fetal circulations (MacMahon et al., 1965).

6. Nondisease Effects. Air pollution afflicts us in ways that are intensely disconcerting, even dangerous, but not directly damaging to health. Particulates create haze, reduce visibility, and increase the hazards of traffic on the ground and in the air (Anderson, 1967). Quite often, pollution is malodorous. It can irritate the conjuntival lining of the eyes and the mucosal lining of the throat. Eye irritation is reportedly the most wide-spread symptom of the photochemical pollution in Los Angeles County; it is experienced by about three-fourths of the population (Goldsmith, 1969).

We are only just beginning to study the possible impact of air pollution on behavior (Anderson, 1967). We surmise that aesthetic reactions are among those driving the more affluent segments of society out of polluted cities. We—probably most of us—are frustrated and angered by the sight of a city or a valley enshrouded in dirt.

How strong is the case that urban air pollution affects health? Most authorities would probably agree that the evidence is good and that it is being continually strengthened, though gaps and weaknesses persist. It is unlikely that we shall ever know precisely all the effects on man of individual pollutants or their mixtures. Both the population and the pollution are too dynamic, too changing, and ultimately too difficult to characterize to permit unqualified answers. Perhaps for this reason, we ought to upgrade our reliance on the secondary criteria concerned with welfare, and give more emphasis to aesthetic and economic factors.[7] This suggestion is not intended to demean our concern about health, but rather to broaden the basis for our decisions and to take into fuller account the enormity of the problem.

[7]Lave and Seskin (1970) estimated that $2,080 million would be saved annually by a 50% reduction in levels of air pollution.

REFERENCES

Amdur, M. O., 1961. The effect of aerosols on the response to irritant gases. In *Inhaled particles and vapours,* ed. C. N. Davies. New York: Permagon.

Anderson, D. O., 1967. The effects of air contamination on health; a review. Part II. *Can. Med. Assoc. J.* 97:585-593.

Bates, D. V., et al. Short-term effects of ozone on the lung. *J. Appl. Physiol.,* in press.

Buell, G. C., et al., 1965. Potential crosslinking agents in lung tissue. *Arch. Environ. Health* 10:213.

Carnow, B. W., 1970. Relationship of SO_2 levels to morbidity and mortality in "high risk" populations. Presented at the Air Pollution Medical Research Conference, American Medical Association, New Orleans, October 5–7.

———, et al., 1969. Chicago air pollution study. SO_2 levels and acute illness in patients with chronic bronchopulmonary disease. *Arch. Environ. Health* 18:768.

Coffin, D. L., and E. J. Blommer, 1965. The influence of cold on mortality from streptococci following ozone exposure. *J. Air Poll. Control. Adm.* 15:523.

———, et al., 1968. Influence of ozone on pulmonary cells. *Arch. Environ. Health* 16:633.

Denton, R., et al., 1965. Chemical engineering aspects of obstructive lung disease. *Chem. Eng. Medicine* 62:12–18.

Douglas, J. W. B., and R. E. Waller, 1966. Air pollution and respiratory in-in children. *Brit. J. Prevent. Social Med.* 20:1.

DuBos, R., et al., 1968. Lasting biological effects of early environmental influences. I. Conditioning of adult size by prenatal and postnatal nutrition. *J. Explt. Med.* 127:783.

Ehrlich, R., and M. C. Henry, 1968. Chronic toxicity of nitrogen dioxide I. Effect on resistance to bacterial pneumonia. *Arch. Environ. Health* 17:860.

Environmental Protection Agency, 1971. *Air quality criteria for nitrogen oxides.* Washington, D.C.

Frank, N. R., et al., 1969. SO_2 (^{35}S Labeled) Absorption by the nose and mouth under conditions of varying concentration and flow. *Arch. Environ. Health* 18:315.

Freeman, G., et al., 1968. Environmental factors in emphysema and a model system with NO_2. *Yale J. Biol. Med.* 40:5, 6, 566.

Goldsmith, J. R., 1969. Nondisease effects of air pollution. *Environ. Res.* 2:93.

———, et al., 1969. Epidemiologic appraisal of carbon monoxide effects, Committee on Effects of Atmospheric Contaminants on Human Health and Welfare (NAS and NAE): *Effects of chronic exposure to low levels of carbon monoxide on human health, behavior, and performance.* Washington, D.C.

Henry, M. C., et al., 1970. Chronic toxicity of NO_2 in squirrel monkeys III.

Effect on resistance to bacterial and viral infection. *Arch. Environ. Health* 20:566.

Higgins, I. T. T., and J. R. McCarroll, 1970. Types, ranges, and methods for classifying human pathophysiologic changes and responses to air pollution. In *Development of air quality standards,* eds. A. Atkisson and R. S. Gaines. Chas. E. Merrill: Columbus, Ohio.

Holland, W. W., et al., 1969. Factors influencing the onset of chronic respiratory disease. *Brit. Med. J.* 2:205.

Lave, L. B., and E. P. Seskin, 1970. Air pollution and human health. *Science* 169:723.

Lawther, P. J., 1966. Air pollution, bronchitis and lung cancer. *Postgrad. Med. J.* 42:703.

MacMahon, B., et al., 1965. Infant weight and parental smoking habits. *Am. J. Epidem.* 82:247.

National Academy of Sciences and National Academy of Engineering, 1969. Effects of chronic exposure to low levels of carbon monoxide on human health, behavior, and performance. Washington, D.C.

National Air Pollution Control Administration (NAPCA), 1969a. *Air quality criteria for sulfur oxides,* summary and conclusions. Arlington, Va.

———, 1969b. *Air quality criteria for particulate matter,* summary and conclusions. Arlington, Va.

———, 1970a. *Air quality criteria for photochemical oxidants,* summary and conclusions. Washington, D.C.

———, 1970b. *Air quality criteria for hydrocarbons,* summary and conclusions. Washington, D.C.

———, 1970c. *Air quality criteria for carbon monoxide,* summary and conclusions, Washington, D.C.

National Institute of Environmental Health Sciences, 1970. Task Force on Research Planning in Environmental Health Science. Man's health and the environment—some research needs. Washington, D.C.

GENE E. LIKENS, F. HERBERT BORMANN, AND NOYE M. JOHNSON

Acid Rain

INTRODUCTION

Rain in New England is becoming increasingly acidic. Acid levels in precipitation now are ten to one hundred times higher than normal. The trend may be linked to air pollutants such as sulfur and nitrogen oxides that can be converted chemically in the atmosphere into strong acids that are then dissolved in rain or snow and fall to earth. The magnitude of the situation already approaches that in the Scandinavian countries. There rain and snow with high acid levels associated with air pollution from industry have been blamed for reduced forest growth, interference with aquatic life, and expensive corrosion damage to buildings.

The accompanying article by Cornell, Yale, and Dartmouth university scientists documents rising acidity in precipitation in four northeastern states: New York, New Hampshire, Massachusetts, and Connecticut. A news release from Cornell University during preparation of the article reports that acid rain is also falling in Maine, Vermont, and Rhode Island.

The implications for human health are not clear-cut. Extremely acid water would be hazardous for bathing or drinking, but rainwater that drains from land into waterways generally becomes neutralized by contact with minerals in the soil and thus would not normally increase the acidity of water used for swimming or drinking. Acid rain that falls directly into waterways often would be made less so by compounds already naturally present in the water. In fact, extreme acidity in drinking supplies is so rare that the official U.S. Public Health Drinking Water Standards do not bother to set limits on acidity or alkalinity. The Cornell University news release, however, prompted *New York Times* reporter David Bird to check with an official of the

New York City Department of Water Resources. The official reported that there has been no discernible increase in the acid level of the New York City water supply over the years, even if there has been increased acidity of precipitation in the Northeast.

Aside from its potential direct effect on bodies of water used by man, acid rain caused by air pollution may indirectly signal other dangers. One threat may be to man's respiratory system. Sulfur dioxide is one of the pollutants which is implicated in the acid trend reported by Likens and his colleagues in the accompanying article. Sulfur dioxide in the atmosphere is partially converted to sulfur trioxide, which is immediately converted to sulfuric acid in the presence of water. Sulfuric acid is more irritating to the lungs than is sulfur dioxide, particularly if the acid is suspended in a fine mist of the sort encountered in London, according to the U.S. Public Health Service Air Quality Criteria for Sulfur Oxides. Sulfuric acid in particles as large as raindrops probably would not pose a direct threat to the respiratory system, but heavy concentrations of the same acid in small particles of water on a misty day could present such a danger. The point is that the acid trends discovered by Likens and his fellow scientists serve as a serious warning that a massive change in environmental conditions is occurring which must be assessed in broadest terms of human and ecological well-being.

ACID RAIN

European scientists have found that rain in northwestern Europe shows a trend toward increased acidity, particularly over the past fifteen years. The tendency appears linked to mounting levels of certain gaseous pollutants such as sulfur and nitrogen oxides, which can be converted chemically in the atmosphere to strong acids. Although the trend appears to pose no apparent threat to health, it can do considerable damage to man-made structures and equipment, and more importantly it has serious implications for ecological systems.

Data indicating this trend are expressed in what are called pH values. These values are a measure of hydrogen ion activity. Briefly, water slightly ionizes (is converted into ions, which are electrically charged particles), yielding hydrogen and hydroxyl ions. When the activity of these ions is equal, water is neither acid nor alkaline and is said to be neutral. This neutrality point is represented by the pH value 7, which actually varies slightly according to temperature. Water becomes increasingly acid at pH values below 7, increasingly alkaline at values above 7. There are tenfold differences between each unit. Thus, pH 6 is ten times more acid than pH 7, pH 5 is 100 times more acidic than pH 7, and so on.

Distilled water in contact with carbon dioxide in the air becomes slightly acid since carbon dioxide combines with water to form carbonic acid. Normally, the pH of rainwater, regulated only by carbon dioxide gas in the atmosphere, would be 5.7 or only slightly acid (Barrett and Brodin, 1955). Recent European studies, however, have reported increasing quantities of much stronger acids in rain and snow, producing pH values between 3 and 5. The presence of these acids is presumably related to air pollution.

By burning fossil fuels, man releases large quantities of sulfur and nitrogen oxides to the atmosphere. These compounds are in addition to naturally occurring gaseous forms of sulfur and nitrogen. It is estimated that about 3,740,000 tons of sulfur are released to the atmosphere each year by the combustion of fossil fuels (Berntine, 1971). In 1968 man-made emissions from the U.S. contributed some 33.2 million tons of sulfur oxides and 20.7 million tons of nitrogen oxides to the atmosphere (MIT, 1970). Sulfur dioxide and hydrogen sulfide are chemically changed (oxidized and hydrolyzed) in the atmosphere to sulfuric acid at varying rates depending upon environmental conditions. Likewise, various nitrogen oxides are transformed into nitric acid. If these acids are not neutralized by alkaline substances also present in the atmosphere, they will ultimately fall to the land and waters in precipitation.

With unpolluted atmospheric conditions, the relatively weak acidity of carbonic acid in precipitation is one of the major factors in the chemical weathering of landscapes. (Weathering is the decomposition of rocks.) The chemical weathering process is in part responsible for the slow but steady leaching (removal) of nutrients from primary and secondary minerals in rocks and soils. By changing the acidity of precipitation, man's activities may have important modifying effects on these and other vital functions of natural ecosystems. Very little attention has been given to this problem in the literature.

Acid rain is not a new phenomenon although it may now be affecting large geographic regions. For example, Crowther and Rustan report that substantial amounts of acid were added to the soil in the industrial city of Leeds, England, in rainfall during 1907–1908 (Crowther, 1911). Expressed as sulfuric acid, this amounted to 1.84 pounds of hydrogen ions per acre per year (sulfuric acid dissociates in water to release hydrogen and sulfate ions). By way of reference, one meter (about 40 inches) of rain per year at a pH of 4 would add 2.20 pounds of hydrogen ions per hectare (2.47 acres). At a pH of 5, the input would be 0.22 pounds per hectare, and at pH 3 it would be 22 pounds per hectare. The rain in Leeds was thus strongly acidic.

Increasing acidification of precipitation and the consequent ecological effects have received the most attention in Scandinavia (Oden, 1968,

1970; Lundholm, 1970; Reiquam, 1970; Engstrom, 1971). There it is estimated that more than 75% of the sulfur in the air is produced by human activity. Much of this sulfur apparently comes from distant industrialized regions, such as England and the Ruhr Valley. The average residence time for sulfur in the atmosphere is two to four days, and on the average it may be transported more than 620 miles before being deposited on the earth's surface (Engstrom, 1971). Because of the availability of data from a network of sampling stations in Scandinavia, it is possible to map changes in acidity of precipitation in recent years. (See Figure 1.) The results show that there has been a striking increase, with the acidity of rain in some parts of Scandinavia increasing more than 200-fold since 1956 (Oden, 1970). Values in rain as low as pH 2.8 have been recorded (Oden, pers. comm.).

Gorham has observed the relation between air pollution and the acidity of precipitation in the English Lake District, where the pH of rain averages less than 4.5 and occasionally may be less than 4.0 (Gorham, 1958). Holden reports mean annual pH values of 5.08 to 4.74 to 4.40 in a southerly direction for three inland Scottish locations (Holden, 1966), corresponding to increasing excess sulfate values and industrial atmospheric pollution. Even rain collected near Manaus in the center of the Amazon basin had pH values ranging between 4.0 and 5.4 during 1966–1968 (Ungemach, 1970); these values may be related to large-scale burning of forests in this area (Forman, pers. comm.). A few pH values for other localities in the world have been summarized (Carroll, 1962).

We first became aware of the problem of acid rain through our studies of biogeochemical cycles for the Hubbard Brook Experimental Forest in New Hampshire. Continuous measurement of pH in rain and snow samples since 1964 has clearly shown that precipitation in this rural, forested area is quite acid (Fisher, 1968). In fact, the hydrogen ion is the predominant positively charged ion in precipitation, indicating the presence of strong acids such as sulfuric and nitric acids. The annual weighted average pH between 1965 and 1971 ranged between 4.03 and 4.19. (A weighted average takes into account the amount of rain as well as its composition.) Although there has been no apparent trend, 1971 is the only year in which the pH was less than 4.1. The lowest pH value recorded at Hubbard Brook was a surprisingly acid 3.0.

How acid is rainfall over the United States? Is the acidity changing and why? These questions are vital to considerations of environmental quality, but unfortunately answers are based on scanty information. However, from the few data that are available, it would appear that the acid precipitation is much more widespread than generally believed. We have attempted to assemble data to characterize the acidity of rain

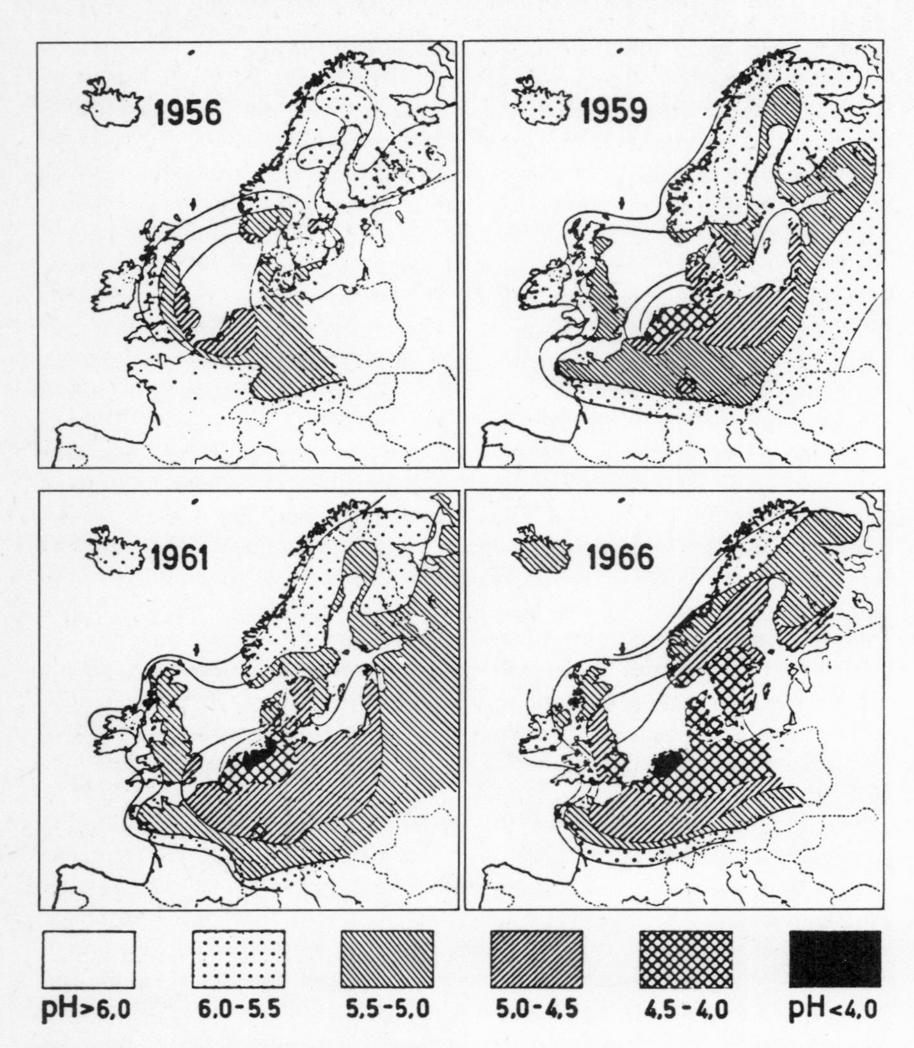

Figure 1 Acid precipitation in Europe
Changes in the pH of precipitation over northern Europe from 1956 to 1966. Source: Oden, S., "Nederbordens forsurning-ett generellt hot mot ekosystemem," I. Mysterud (red.), *Forurensning og biologisk miljovern*, Universitetsforlaget, Oslov, 1971, pp. 63–98.

and snow in the U.S., but there have been few comprehensive studies of precipitation chemistry, and pH values often are not detailed. Most seriously, we can find no data to accurately describe long-term trends for the United States.

Measurements recently completed in the central Finger Lakes Region of New York State show that rain and snow are equally as acid as that found in New Hampshire. The annual pH, based upon samples collected during the twelve-month period 1970–1971, and weighted proportionally to the amount of water and pH during each period of precipitation, was 3.98 at Ithaca, New York; 3.91 at Aurora, New York; 4.02 at Geneva, New York; and 4.03 at Hubbard Brook, New Hampshire. Summertime values were lower for all four sites. Because local combustion of fossil fuels is greater during the winter, these variations may reflect different seasonal paths for regional air masses carrying sulfur or nitrogen derivatives from more distant industrialized areas. We have not yet completely analyzed our data in this regard, but it is noteworthy that major storm tracks and air masses generally differ between summer and winter seasons in central New York (Dethier, 1966). Current measurements over the entire state indicate that rain and snow are surprisingly and consistently acid (Table 1). Based upon these data and those of Pearson and Fisher (1971), it is apparent that the precipitation falling on most of the northeastern U.S.

Table 1 Annual Weighted pH for Precipitation in New York State[a] (The Range of Observed Values is Given in Parentheses)

	1965–66[b]	1966–67	1967–68	1968–69
Mineola	—	4.42 (4.0–6.4)	4.45 (3.8–5.2)	4.57 (4.2–6.5)
Upton	(3.9–4.5)	4.46 (4.2–4.9)	4.33 (3.8–5.0)	4.11 (3.9–4.8)
Rock Hill	(4.3–5.6)	4.20 (4.0–4.8)	4.26 (4.0–5.0)	4.29 (4.0–4.8)
Albany	(4.2–4.4)	4.87 (4.3–6.5)	4.61 (4.2–6.6)	4.68 (4.3–6.8)
Hinckley	(4.1–4.6)	4.45 (4.2–6.9)	4.37 (4.0–6.4)	4.37 (4.1–4.7)
Allegany State Pk.	(6.1–6.5)	4.27 (4.0–7.2)	4.53 (4.1–6.0)	4.31 (4.1–7.0)
Mays Point	(4.1–4.3)	4.29 (4.0–4.8)	4.78 (4.2–7.1)	4.93 (4.3–7.0)
Canton	(4.3–4.4)	4.34 (4.0–5.0)	4.57 (4.3–5.9)	4.37 (4.1–5.5)
Athens, Pa.	(4.5–4.6)	4.34 (4.0–6.1)	4.25 (4.0–5.9)	4.28 (4.0–5.4)

[a]Calculated from U.S. Department of the Interior, Geological Survey, Water Resources Data for New York, Part II; Water Quality Records, 1966 through 1969, Albany, New York.
[b]August-October 1965.

is characterized by high acidity. Weighted annual pH values of 4.27, 4.29, and 4.27 have been reported for New Durham, New Hampshire; Hubbardston, Massachusetts; and Thomaston, Connecticut. At Thomaston, the lowest monthly average pH, 3.9, was recorded at the height of the growing season (Fisher, pers. comm.). In New Haven, Connecticut, pH values during the spring and summer of 1970 averaged 3.81, with a summer average of 3.62, and a low of 3.52. Killingworth, located 30 miles east of New Haven, recorded a pH of 4.31 in April 1970.

Although extensive data are available for New York on nitrogen and sulfur in precipitation during the first half of this century (Collision and Mensching, 1932; Wilson, 1926; Pearson and Fisher, 1971), no concurrent published records of pH have been found. Apparently, prior to 1940, when most of this rainwater chemistry was done, only methyl orange was used to indicate acid or alkaline conditions in samples. Unfortunately, this indicator is of little use in describing the pH during this period since methyl orange changes color at a pH of 4.6, meaning that actual determinations of pH above or below this level could not be made. In the absence of data on pH, the large amounts of bicarbonate in rainwater samples taken at Geneva, New York during the period 1919–1929 (Collision and Mensching, 1932), indicate much higher pH values than today. Bicarbonate, from carbonic acid, cannot coexist with the stronger acids found in today's rain. The presence of bicarbonate, therefore, would indicate that pH values in 1919–1929 were 5.7 or higher.

Available data on the pH of rain and snow in other parts of the U.S. help to define the problem. The acidity of rain and snow in the Pacific and eastern Gulf coastal regions of the U.S. is of particular interest in evaluating the localized effect of industrial air pollution, since prevailing air flow is onshore. The pH of rainfall at Corvallis, Oregon, in December 1969 and January 1970 ranged between 4.95 and 5.7 (Malneg, pers. comm.). Tarrant and his colleagues found that the pH of precipitation ranged between 5.7 and 6.3 (annual average of 6.1) during the period 1963–1964 in coastal Oregon forests (Tarrant et al., 1968). The pH of glacial ice in the northern Cascade Mountains, which represents a historical composite of snow, has been measured at 5.6 (Reynolds and Johnson, 1972). Rainwater values ranging from pH 5.4 to 8.5 in 1957–1958 and pH 4.9 to 6.8 in 1958–1959 at Menlo Park, California, have been reported (Whitehead and Feth, 1964). Brezonik and his colleagues observed pH values between 5.3 and 6.8 during 1967–1968 in north central Florida (Brezonik). Collectively, these pH values suggest a carbonic acid control for the precipitation of these areas, which by and large are meteorologically isolated from large industrial centers.

(Carbonic acid control refers to acidity due to the carbonic acid r
than to stronger acids such as sulfuric acid.)

THE EFFECTS ON ECOSYSTEMS

Simply stated, the problem is that water from rain and snow in many areas of the world is no longer characterized by the weak, although highly buffered (containing substances that prevent change in pH of a solution after the addition of either acid or alkali), carbonic acid, but instead is dominated by the stronger, although unbuffered, sulfuric and nitric acids, at a greatly reduced pH. The ecological effects of this change are as yet unknown, but potentially they are manifold and very complex. Effects may range from changes in leaching rates of nutrients from plant foliage, changes in leaching rates of soil nutrients, acidification of lakes and rivers, effects on metabolism of organisms, and corrosion of structures. These effects illustrate the dynamic and complex linkages between ecosystems.

Since most combustion of fossil fuels occurs in the mid-latitudes of the Northern Hemisphere, any consequent changes in acidity will be most evident here. However, there are separate and synergistic (combined) effects of other atmospheric and land-based pollutants such as sulfur dioxide, nitrogen oxides, and ozone. Thus, the effects of acid precipitation in ecosystems are often difficult to isolate from other man-induced changes that are also accelerating. The best data come from Scandinavia.

A trend of decreasing pH (increasing acidity) was observed during 1965–1970 in almost all lakes and rivers covered by a network of sampling stations in Scandinavia (Oden and Ahl, 1970). Acidity increased at a rate of 8 to 24% per year in Swedish rivers. This corresponds to the period of increasing acidity of rainfall for this area (Fig. 1), but also must represent the effects of other pollutants dumped into aquatic ecosystems by man's activities. Significantly, unpolluted river systems also increased in acidity (in five years the pH was lowered 0.15 units, that is, the acidity increased 1.4 times); these changes were attributed directly to the acidification of precipitation.

The effects of increasing acidity on aquatic life may be very serious. According to Grande and Snekvik: "This fallout, which is particularly high in sulfur dioxide, has already so altered the pH of certain streams in southern Norway that salmon eggs can no longer develop and the salmon runs have been eliminated," (Grande and Snekvik, 1971). In most unpolluted, fresh water, the pH generally ranges between 6.5 and 8.5 (FWPCA, 1968).

Effects on soil fertility and biological productivity are intertwined and difficult to evaluate. Increased input of acid into soil may lead to increased leaching of calcium and other nutrient minerals. Such losses may not result in any significant short-term damage to arable land (land capable of producing crops), but they represent an added stress to the ecosystem. If it became necessary for man to replace chemicals lost from the soil as a result of acidification of rain, very large costs could be involved. In Sweden, it is thought that in the last fifteen years there has been an annual reduction in forest growth as a result of acid rainfall (Engstrom, 1971).

Thus, these data indicate fundamental changes in land and water ecosystems resulting from acidified rainfall. This same phenomenon must be occurring in other parts of the world, but there is little cognizance of it, and it is often confounded by other sources of pollution. The eastern U.S., particularly New England, undoubtedly receives sulfur and nitrogen oxides from the West and Midwest, but large-scale changes and effects associated with increased inputs of these compounds and increased acidity of preciptation have been largely overlooked or ignored. Long-term records of water quality in lakes and rivers of the U.S. are scarce and difficult to interpret (Wolman, 1971). Similarly, any effects of acid rain on these ecosystems is at present unclear, but a few long-term observations are interesting and suggestive. Schofield (1965) reports that the water chemistry of a relatively large clear-water, oligotrophic (deficient in plant nutrients and low in productivity) lake in the Adirondack Mountains of New York has changed appreciably since 1938. In December 1938, the alkalinity expressed as chemically equivalent concentrations of calcium carbonate ranged from 12.5 to 20.0 parts per million and the pH was 6.6 to 7.2, whereas during 1959–1960 the alkalinity ranged from 0 to 3.0 parts per million and the pH ranged from 3.9 to 5.8, demonstrating a sharp increase in acidity. Although the information prior to 1910 is scanty, it would appear that there was a significant jump in the sulfate content of Lake Michigan, the Illinois River, and possibly the Ohio River by 1920–1930, with a more gradual increase in concentration to the present time (Ackerman et al., 1970). Significantly, in this regard, alkalinity values in Lake Michigan have decreased about 7% during the past 60 years, suggesting increased acidity. A similar decrease in alkalinity of about 2% in the Mississippi River was observed between 1911 and 1968. The pattern in the Illinois and Ohio rivers is more complex, with periods of increase and decrease since the early 1900s. The interpretation of such data is very difficult, but changes in all of the environmental inputs must be considered and evaluated in assessing long-term fluctuations in environmental quality.

Sweden's case study for the UN Conference on Human Environment, held in Stockholm in the summer of 1972, deals with "air pollution across national boundaries." The early results clearly indicate that air pollution is an unpremediated form of chemical warfare! Data collected on atmospheric pollutants and acid rain suggest a very serious problem in northern Europe. Existing data suggest that the problem in the U.S., particularly in the Northeast, has already reached similar proportions. We urge consideration of these data in the establishment of air pollution standards and a massive effort to increase our understanding of this problem. Detailed ecological and geochemical studies are urgently needed on the separate and combined effects of additions of nitrogen and sulfur oxides and associated acidity on ecosystems.

Acknowledgements

This is contribution No. 48 of the Hubbard Brook Ecosystem Study. Financial support for the field study at Hubbard Brook was provided by the National Science Foundation, and was done through the cooperation of the Northeastern Forest Experiment Station, Forest Service, U.S. Department of Agriculture, Upper Darby, Pennsylvania. Support for studies of precipitation chemistry in the Finger Lakes Region of New York was provided by the U.S. Department of Interior, Office of Water Resources Research through the Cornell University Water Resources and Marine Sciences Center. Climatological data for New York were obtained from the U.S. Department of Commerce, Environmental Sciences Services Administration. We thank John S. Eaton, Cornell University, for computational assistance and Ray T. Oglesby, Cornell University, and C. A. Federer, U.S. Forest Service, for comments on the manuscript.

REFERENCES

Ackerman, W. C., et al., 1970. Some long-term trends in water quality of rivers and lakes. *Trans. Am. Geophys. Union* 51:516–522.

Barrett, E., and G. Brodin, 1955. The acidity of scandinavian precipitation. *Tellus* 7:251–257.

Bertine, K. K., and E. D. Goldberg, 1971. Fossil fuel combustion and the major sedimentary cycle. *Science* 173:233–235.

Brezonik, P. L., et al., *Eutrophication factors in north central Florida lakes.* Florda Engineering and Industrial Expt. Station, Gainesville, Bull. Series 134, Water Resources Research Center Publ. No. 5.

Carroll, D., 1962. Rainwater as a chemical agent of geological processes—a review. Geol. Surv. Water-Supply Paper 1535-G, 18 pp.

Collision, R. C., and J. E. Mensching, 1932. Lysimeter investigations: II. Composition of rainwater at Geneva, N.Y. for a 10-year period. New York Agric. Expt. Station, Tech. Bull. No. 193:3–19.

Crowther, C., and H. G. Rustan, 1911. The nature, distribution and effects upon vegetation of atmospheric impurities in and near an industrial town. *J. Agr. Sci.* 4:25–55.

Dethier, B. E., 1966. Precipitation in New York State. Cornell University Agr. Expt. Station, Ithaca, N.Y., Bull. No. 1009, 78 pp.

Engstrom, A., 1971. *Air pollution across national boundaries, the impact on the environment of sulfur in air and precipitation.* Report of the Swedish Preparatory Committee for the U.N. Conference on Human Environment, Kungl. Boktryckeriet P.A. Norstedt et Soner, Stockholm, 96 pp.

Federal Water Pollution Control Administration, 1968. *Water quality criteria.* Report of the National Technical Advisory Committee of the Secretary of the Interior, Washington D.C., 234 pp.

Fisher, D. W., et al., 1968. Atmospheric contributions to water quality of streams in the Hubbard Brook Experimental Forest, New Hampshire. *Water Resources Res.* 4:1115–1126.

Fisher, D. W., personal communication.

Forman, R., personal communication.

Gorham, E., 1958. Free acid in British soils. *Nature* 191:106.

Grande, M., and E. Snekvik, 1971. *Major pollution problems affecting inland fisheries in Norway.* Report of a Meeting of Representatives of the Norwegian Water and Hydro-Electricity Board, the Farmer's Association, The Institute for Water Research, and the Fish and Wildlife Service, Oslo, February 26, 1969 (quoted from Klein, D. R., Reaction of reindeer to obstructions and disturbances. *Science,* 173:393–398).

Holden, A. V., 1966. A chemical study of rain and stream waters in the Scottish Highlands. *Freshwater Salmon Fisheries Res.* 37:3–17. Dept. of Agriculture and Fisheries, Edinburgh, Scotland.

Junge, C. E., 1963. *Air chemistry and radioactivity.* New York: Academic Press.

Leland, E. W., 1952. Nitrogen and sulfur in the precipitation at Ithaca, N.Y. *Agron J.* 44, no. 4:172–175.

Lodge, J. R., Jr., et al., 1968. *Chemistry of United States precipitation.* Final Report on the National Precipitation Sampling Network, National Center for Atmospheric Research, Boulder, Colorado, 66 pp.

Lundholm, B., 1970. Interactions between oceans and terrestrial ecosystems. In *Global effects of environmental pollution,* ed. S. F. Singer, pp. 195–201. New York: Springer-Verlag.

Malueg, K., personal communication.

Massachusetts Institute of Technology, 1970. *Man's impact on the global environment: report of the study of critical environmental problems.* Cambridge, Mass.: MIT Press.

Oden, S., 1968. Nederbördens och Luftens Försurning-dess Orsaker, Förlopp och Verkan I Olida Miljöer, Statens Naturvetenskapliga Forskningsrad, Stockholm, Bull. No. 1, 86 pp.

Oden, S., and T. Ahl, 1970. *Försurningen av skandinaviska vatten (The acidification of Scandinavian lakes and rivers).* Ymer, Arsbok, pp. 103–122.

Oden, S., personal communications.

Pearson, F. J., Jr., and D. W. Fisher, 1971. Chemical composition of atmospheric precipitation in the northeastern United States, Geol. Surv.—Supply Paper 1535-P, 23 pp.

Reiquam, H., 1970. European interest in acidic precipitation. In *Precipitation scavenging,* eds. R. J. Engelmann and W. G. N. Slinn. U.S. Atomic Energy Symposium, Series 22, Division of Technical Information, Oak Ridge, Tennessee, pp. 289–292.

Reynolds, R. C. and N. M. Johnson, 1972. Chemical weathering in the temperate glacial environment of the northern Cascade Mountains, *Geochim. Cosmochim. Acta,* in press.

Schofield, C. L., Jr., 1965. Water quality in relation to survival of brook trout, *Salvelinus fontinalis* (Mitchill). *Trans. Am. Fisheries Soc.* 94, no. 3:227–235.

Tarrant, R. F., et al., 1968. Nutrient cycling by throughfall and stemflow precipitation in three coastal Oregon forest types. USDA Forest Service, Research Paper PNW-54, Pacific Northwest Forest and Range Expt. Station, Portland, Oregon, 7 pp.

Ungemach, H., 1970. Chemical rain water studies in the Amazon region. In *Il simposio y foro de biologia tropical Amazonica,* ed. J. M. Idrobo. Bogata, Columbia: Editorial Pax.

Whitehead, H. C., and J. H. Feth, 1964. Chemical composition of rain, dry fallout, and bulk precipitation at Menlo Park, California 1957–1959. *J. Geophys. Res.,* 69, no. 16:3319–3333.

Wilson, B. D., 1926. Nitrogen and sulfur in rainwater in New York. *J. Am. Soc. Agron.* 18:1108–1112.

Wolman, M. G., 1971. The nation's rivers. *Science* 174:905–918.

MICHAEL FORTUNE

Environmental Consequences of Extracting Coal

UNDERGROUND MINING

The bulk of the nation's coal originates from underground mines throughout Appalachia. This method, although less disruptive to the earth than strip mining, results in higher prices and exacts a larger toll of human life. The environmental effects of concern are acid mine drainage, subsidence of the overburden, processing wastes, and long-lasting fires.

Four million tons of unneutralized sulfuric acid seeping from coal mines annually degrade about 10,500 miles of Applachian streams and kill an estimated one million fish (FWQA, 1970). Actually twice this amount seeps from the mines, but one half is neutralized by natural alkalinity in the receiving waters. Ground water and free oxygen react with the sulfur naturally occurring in coal; the resultant acid dissolves iron, aluminum, manganese, and other minerals before it seeps to the surface. Small amounts of acid can severely limit biological productivity and eliminate the less resistant species; if the pH drops below 5 fish cannot survive. Associated with acid drainage and with the lime that may be used to neutralize it is "yellow boy," a sludge of ferric hydroxide which accumulates on structures and stream beds, smothering life and reducing the capability of gravel beds to percolate oxygen through the water. The Federal Water Quality Administration estimates that mine drainage costs Appalachia $7.5 million yearly in recreation income alone. It can be effectively prevented only by designing the entrance to the mine so that it lies above the groundwater table.

Nearly as serious is the subsidence of the rock strata overlying abandoned coal mines. The Department of the Interior calculates that of the eight million acres that have been undermined, two million acres have subsided; 7% of this lies in urban areas (Aaronson, 1970). Rural land subsidence alters drainage patterns, damages crops and depresses land values; in the city it ruptures pipes and causes buildings to collapse. The mine operator may deliberately "rob"' the thick pillars of coal as he retreats before the collapsing ceiling, or explosions, fires, or mere exposure to air may weaken the pillars. Pennsylvania is the only state that makes the operator responsible for subsequent damages. The settling may be prevented by backfilling the mine with coal processing wastes or by inserting permanent steel roof supports.

Coal processing results in a number of wastes of various reactivity. 93% of the coal processed is washed with water, a process which discharges "black-water" containing mostly inert dust and silt; this can easily be filtered or settled. Dissolved chemicals including cyanide, hydrochloric, and sulfuric acid pose more difficulty; treatment ranges from 10% to 99% effective at costs from 10¢ to $3.25 per thousand gallons. These wastes are increasing at such a rate that only a 95% effective treatment will reverse the trend (Aaronson, 1970).

The processing plants have accumulated 2.2 billion tons of solid wastes up to 1966; this amount should be *annually* produced in the year 2000. These wastes are piled in unstable banks that occasionally collapse as one recently did in Aberfan, Wales, killing 150 children. If a refuse bank catches fire it can be notoriously difficult to extinguish. Five hundred are presently burning in the United States and emitting carbon monoxide, hydrogen sulfide, and sulfur oxides; 60% of these are within one mile of a community. The best methods for avoiding such fires are to compact the banks and seal them with clay.

Two hundred ninety-two fires burning in underground coal seams deplete our reserves, cause subsidence and emit noxious gasses to the surface. Underground fires are also difficult to extinguish: one has burned at New Straightsville, Ohio, since 1884, despite over one million dollars spent trying to put it out; $5.5 million has been spent so far in vain on a fire in Scranton, Pennsylvania. Again, foresight is the only cure—65% of these blazes start in garbage heaps on abandoned strip mines.

STRIP MINING FOR COAL

"Strip mining is like taking seven or eight stiff drinks. You are riding high as long as the coal lasts, but the hangover comes when the coal is gone, the land is gone, the jobs are gone, and the bitter truth of the morning after leaves a barren landscape and a mouth full of ashes."

—Congressman Ken Heckler, West Virginia

"I like to lose my mind over it."

—Mrs. Bige Ritchie, when she saw a stripping bulldozer uproot the coffin of her infant son and pitch it down the mountainside.

"It has become an Appalachian Carthage, the beginning of a New-World Sahara."

—Harry Caudill

"Some coal operators make an attempt to reclaim land for stripping. But is it enough? Can they restore rolling fields of grain? Can they replace homes that have been deserted and torn down and replaced by acres of dust and stone?"

—Nancy Hudson

"Strip mining is part of the American way."

"Strip mining, while it is going on, looks like the devil, but . . . if you look at what these mountains were doing before this stripping, they were just growing trees that were not even being harvested!"

—James Curry, Reclamation Director,
Tennessee Valley Authority.

Since WW II the more disruptive practice of strip mining has become feasible on a large scale. Currently 40% of the nation's total production of 600 million short tons of coal per year originates from strip mines and the proportion is increasing rapidly. Yet the reserves have barely been scratched—the U.S. Geological Survey estimates that only 4.4 billion tons of coal have been extracted from the 128 billion tons of strippable reserves—about a 240 year supply remaining at present rates (Averitt, 1970). Yet this seemingly large bank of "strippable resources" represents only 8% of the nation's total reserves of coal, 1560 billion tons (Averitt, 1970), which means that we are stripping our coal resources much more readily than we are mining them via underground tunnels. The best reason for this is an economic one—in 1968 the average cost of coal (including royalties) from underground mines was $5.22, while that from strip mines was $3.75. The recent rise in the price of coal combined with the latest UMW wage settlements and the 1969 Federal Mine Safety Act has increased the relative profitability of the strip-mined coal. The basic cost advantage of strip-mined coal lies in the higher output per man-day (100% greater than in underground operations) and a higher recovery rate (60% higher than in underground operations).

Enormous earth-moving machines help to maintain the economic advantage of strip mining. "Big Muskie," the world's largest earth remover with a 220 cubic-yard bucket, is able to scoop 325 tons of earth in one bite and can economically excavate overburden to a depth of 185 feet without resiting itself. Trucks up to 200 tons capacity also expedite the extraction of coal. Such large machinery has permitted a steady increase in the economically feasible ratio of overburden depth to coal seam thickness. Table 1 indicates the trend of the average ratio over the last 25 years. Mines with much larger ratios have been profitably worked however. For example, in Kansas and Illinois the ratio is 30:1. The scale of the mining operations is difficult to comprehend. A proposed strip mine in southwestern Wyoming will measure one mile square and have a highwall 1600–2000 feet high (Averitt, 1970).

Figure 1 shows that the locus of intensive stripping is moving west to Illinois; more recently stripping is gaining strength on Indian reservations in the far West. In 1968 six states accounted for the bulk of stripped-coal production. They are in descending order: Illinois, Kentucky, Ohio, Pennsylvania, Indiana, and West Virginia. Although new areas are being opened, contour mining in the mountains of West Virginia and Pennsylvania has passed its peak and area mining in the flat plains and plateaus of the Ohio Valley, Midwest, and far West has gain tremendously.

A survey of strip-mined land by the U.S. Department of the Interior (1967) in 1965–66 disclosed that 3.2 million acres (5000 square miles) of the nation had been disturbed by strip mining by the beginning of 1965. Coal mining accounted for 41%, or 1,300,000 acres, of the disturbed area. On the basis of production since then, about 1,570,000

Table 1 Average and Maximum Thickness (in feet) of Overburden Removed and Average Thickness of Bituminous Coal and Lignite Recovered by Strip Mining in the United States for Selected Years

	1946	1950	1955	1960	1965	1970 (estimated)
Average thickness of overburden removed	32	39	42	46	50	55
Maximum thickness of overburden removed	—	—	70+	100	125	185
Average thickness of coal recovered	5.2	5.1	4.9	5.1	5.2	5.0
Ratio of average overburden thickness to average coal thickness	6:1	8:1	8.5:1	9:1	10:1	11:1

Source: Young (1967) in Averitt (1970, p. 18)

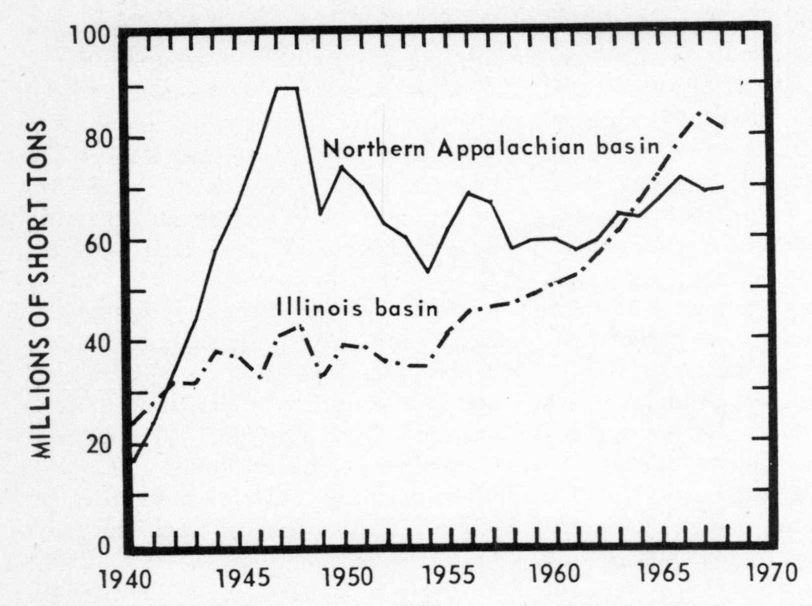

Figure 1 Strip-coal production of the northern Appalachian basin (Pennsylvania, Ohio, West Virginia, and Maryland) versus that of the Illinois basin (Illinois, Indiana, and western Kentucky). Source: U.S. Bureau of Mines (1940-1968).

acres should have been stripped for coal by January 1, 1970. "Disturbed land" is here narrowly defined as the excavation plus the land covered by waste or spoil. It does not include river valleys laden with sediment or acid, summits isolated by highwalls, devalued farmland, or other land adversely affected by the mining. Yet 1.5 million acres is equivalent to one-third of the state of Connecticut.

The Soil Conservation Service judged that only one-third of the total area disturbed by all strip mining had been adequately reclaimed either by natural forces or by deliberate human effort. That leaves 2 million acres of unreclaimed stripped land—only 0.14% of the nation's area, but equivalent to a band of land one mile wide stretching from New York to San Francisco. The Bureau of Sport Fisheries and Wildlife substantiates this figure with their estimate that "two-thirds of the fish and wildlife habitat disturbed could be classified as being severely or moderately affected" (U.S. Dept of Interior, 1967).

The surface coal mine does not resemble the open-pits from which gravel, stone, iron, and copper are extracted. These pits are generally widely scattered, and except for gravel are quite deep, and are characterized by a high ratio of ore to overburden. In contrast the average seam of coal currently being stripped in the U.S. is 5.1 feet thick, and as a rule of thumb ten feet of overburden must be shoveled aside for every foot of thickness of the coal seam. Area mining in flat terrain and contour mining in mountainous regions account about equally for the 1.5 million acres overturned for its coal. In relatively flat terrain a shovel digs a long initial trench down to the seam, then continues to dig new trenches to one side as it fills the old cuts with spoil; eventually a vast area is transformed into a series of ridges up to 50 feet high, spaced 50–100 feet apart, with 17°–39° side slopes. *Area* strip mining such as this is common to Illinois, the Dakotas, and the Southwest.

In mountainous Appalachia contour strip mining is practiced on the coal outcrops on the side of the hills. A shovel cuts a terrace out of the slope where the seam reaches the surface, sometimes completely encircling a mountain. It leaves a nearly vertical "highwall" often 100 feet high where the overburden becomes too thick to profitably remove; some of the loosened rock is piled at the outer edge of the terrace, and the rest is cast down the slope where it may cause landslides and severe erosion. (See Figure 2). On the nation's 25,000 miles

Figure 2 Sources of acid water and silt pollution where the contour method of bituminous coal surface mining is utilized. $

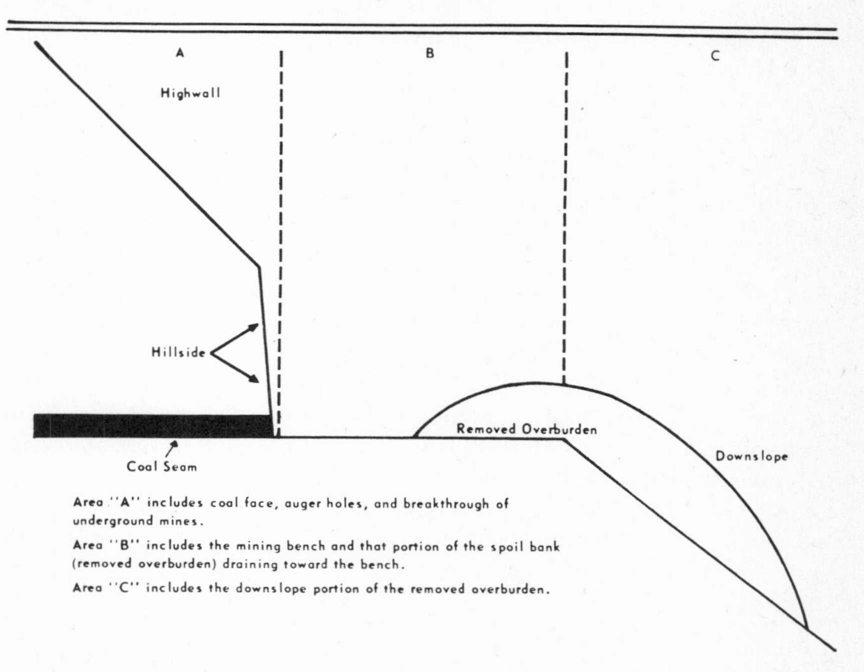

of contour bench, the spoil is stacked on the outer edge for 18,000 miles and pushed entirely down the slope on the remaining 7000 miles. Acidified water often collects in the trench between the highwall and the spoil bank. Contour mining has left 20,000 miles of highwalls that isolate several million acres of summit land, and has caused landslides on 1700 miles of slopes. The Interior Department found evidence (in the form of gullies) of severe erosion at 40% of the sites in their survey, and deposits of sediment in 56% of the ponds and 52% of the streams adjacent to mined areas. Sediment and debris cut the normal storm-carrying capacity of 7000 miles of stream channels. (U.S. Dept. of Interior, 1967).

Surface mining contributes to acid drainage but its share relative to that from underground mines is unresolved. The investigators from the Department of Interior measured a pH less than 5.0 on 48% of the sampled spoil banks in Appalachia, and a pH less than 3.0 on 1% of them. The pH factor is a measure of acidity or basicity. Neutrality is considered 7. The pH of normal Appalachian soil is 6, meaning that it is slightly acidic. The Bureau of Sport Fisheries and Wildlife reported that 5800 miles of streams and 29,000 surface acres of impoundments and reservoirs are seriously affected by surface coal mining. Of the streams receiving direct run-off from the spoil banks, 31% were noticeably laden with precipitates and 37% had discolored water. The precipitate is a sludge of ferric hydroxide (commonly "yellow boy") arising from iron pyrites leached from spoil banks.

The U.S. Geological Survey (Biesecker and George, 1966) conducted perhaps the most comprehensive survey of acid drainage throughout Appalachia in 1965. The water quality at 194 of the 318 sampling sites, a clear majority, was "measurably influenced" by mine drainage—generally by an elevation of sulfate levels. While they did not attempt to identify the actual sources of the acidity, they suggest that surface spoil banks are "limited" contributors to acid drainage because the pyritic material is not continuously in contact with water as it is in underground mines below the zone of saturation.

The Contour Method of strip mining destroys elements of the ecological chain in several ways. The aquatic members die when the water becomes toxic due to acid drainage or the addition of suspended trace metals. Suspended solids are also deposited on the stream bed smothering fauna and flora alike. Non-aquatic members of the chain suffer when the downslope spoil banks are eroded and landslides bury the vegetation. The wildlife habitat has been adversely affected in 13,000 miles of streams, 103,000 acres of lakes, and 1,700,000 acres of land, according to the Interior Department.

The condition of strip-mined land varied widely. Of all the examined spoil banks, 15% supported sufficient vegetation to protect the site, another 15% had "fair to good cover" which should suffice to renew the soil in time, 20% will require seeds and fertilizer, 30% had no cover and will require extensive treatment, and a final 20% cannot support vegetation (and may be impossible to reclaim conventionally) because of excessive stoniness or toxic conditions.

These surveys depict the nationwide damage from all strip mining, including both that for coal and for non-fuel minerals. One gets a more thorough insight into the problem by looking at two case studies—the first of which reveals broad environmental alterations from stripping operations in Beaver Creek basin, Kentucky, and the second discloses socio-economic breakdown in Belmont and Harrison counties, Ohio.

A Case Study—Beaver Creek Basin

The U.S. Geological Survey (1970) conducted an 11-year case study of the environmental effects of contour strip mining in the Beaver Creek basin of Kentucky. They tested two small watersheds, Cane Branch, 10% of which had been stripped from 1955 to 1959, and Helton Branch, a similar region which served as an unstripped control area. The topography was irregular and hilly, with an 8° slope in the stripped area before mining began. The climate was typical of Appalachia—46 inches of mean precipitation, and 22 inches of average annual runoff. The rock strata were siltstone, claystone, and sandstone, covered with pines and oaks on the ridges, and a hardwood and hemlock forest in the coves.

When prospecting began in 1956, the acidity of Cane Branch rose markedly. The basin now yields 1370 tons of dissolved solids per square mile. The spoil banks contribute an astounding 14,000 tons per sq. mi. —126 times the yield of a natural watershed. The waters of the two creeks were chemically analyzed over the years with results in Table 2.

Table 2 Chemical Analysis of Helton Branch and Cane Branch Range of Values and Parts Per Million

Quality	Helton Branch	Cane Branch Before Stripping	Cane Branch After Stripping
Discharge, cubic feet per second	0.12–9.9	—	0.026–52.
Sulfate	0.4–13.	16.	46.–1220.
Iron	0.05–0.15	0.24	0.07–15
Bicarbonate	7.–16.	17.	0
Aluminum	0.1	0.3	1.1–86.
Hardness as $CaCO_3$	7–26	18	34–440

Since 1958 when the most vigorous period of stripping began, the Cane Branch has had an increased flood danger because the peak flow and the rate of change of flow have increased greatly. These changes accompany erosion and sedimentation. The severe erosion from the strip mine has clogged its drainage stream with sediment. Over eight years one part of a spoil bank has eroded at the rate of 14.8 cu. yards per acre, and one gully at an outstanding 159 cu. yards per acre. The gullies have eroded *increasingly* faster with time.

The sediment yield from strip mines is nothing less than astounding. During storms, the sediment concentration in Cane Branch was 30,000 parts per million, compared to 553 ppm in Helton Branch. The yield of sediment from the former basin was 1900 tons per square mile (and 27,000 tons/sq. mi. from the spoil banks) compared to 25 tons/sq. mi. from the latter; the difference between forested and stripped land amounts to a factor of one thousand! Furthermore, the rate of sedimentation is closely tied to the time of mining as Figure 3 shows.

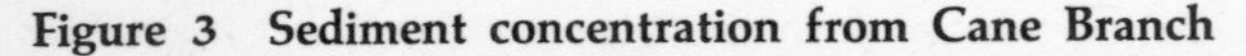
Figure 3 Sediment concentration from Cane Branch

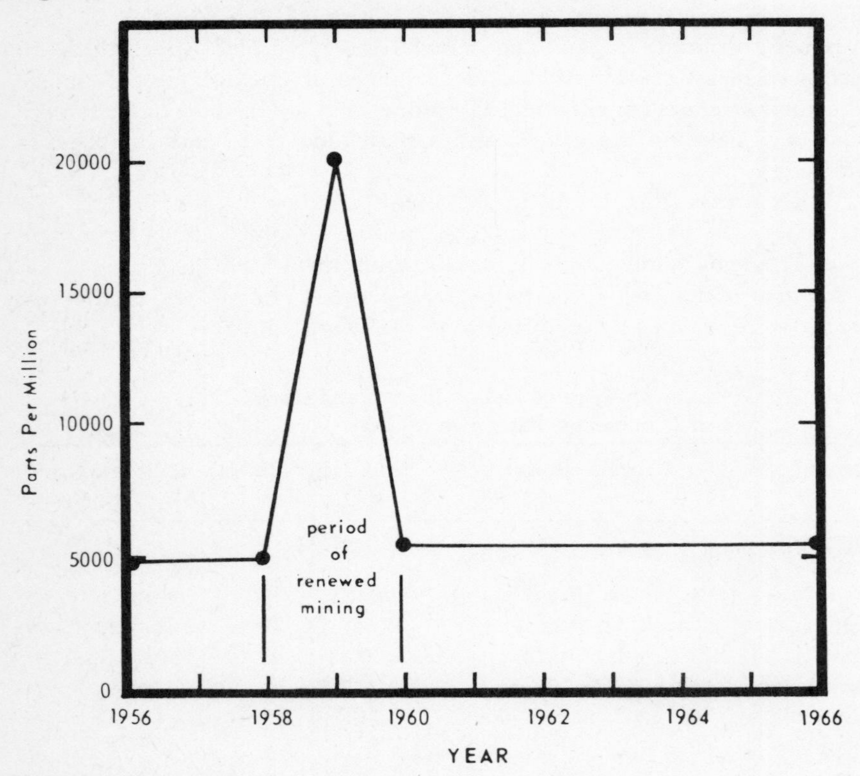

Ninety-eight percent of the sediment in Cane Branch originated in the 10% of its watershed which had been stripped. The growth was poor or nonexistent on 95% of the spoil—not enough to appreciably decrease the sediment yield. One surprising contributor of sediment was the coal haulage roads—their per-acre yields were twice those of the spoil banks.

Sediment, of course, is not innocuous—it deposits itself on the flood plain of the drainage stream and smothers the bottom life. Cane Branch's deposits were 0.5 to 2.8 feet thick.

The ecologists found vast differences between the living populations of the two streams. Cane Branch and its mother stream, Hughes Fork, were devoid of fish. And no wonder—its pH averaged 3 to 4, and the organisms upon which fish feed had been eliminated. Helton Branch, in contrast, enjoyed 5 to 370 pounds per acre of creek chubs and darters.

The fish largely feed on larvae of mayflies and caddisflies, which were absent in Cane Branch but comprised 28% of the total samples in Helton Branch. The abundance and variety of all benthic organisms "differed markedly." Due to thick deposits of silt, clay, and sand, and an unstable stream substrate, Cane Branch had only 30 organisms per *square foot of riffle*[1], compared to Helton's 178.

Even at the bottom of the food chain the differences were remarkable. Whereas Helton had an abundant microflora, the only alga growing in the spoil pools was *Euglena,* and Cane Branch had peculiarly adapted organisms. The bacterium *Ferrobacillus ferrooxidans* lives by oxidizing ferrous iron to ferric iron; filamentous fungi, the yeast *Rhodotorula,* and the alga *Bumillaria* are organisms unique to acid-laden waters such as those in Cane Branch.

Finally, the ecologists noticed that the mine drainage retarded natural reforestation in comparing the biota on the spoil banks to that on adjacent farmland abandoned at the same time as the mining.

Thus the runoff of both dissolved and suspended solids remains exceptionally high and both terrestrial and lotic (stream) ecosystems remain severely degraded at least a decade after strip mining ended.

Case Studies of Belmont and Harrison Counties, Ohio

A southeastern region of Ohio that includes Belmont and Harrison Counties offers a good opportunity to study the socio-economic effects of strip mining. At least two groups have compiled such statistics from the public records of the county governments.

[1]Riffle: the portion of a stream where the gradient is sufficiently steep to cause water to flow or fall over a rocky or gravelly bottom.

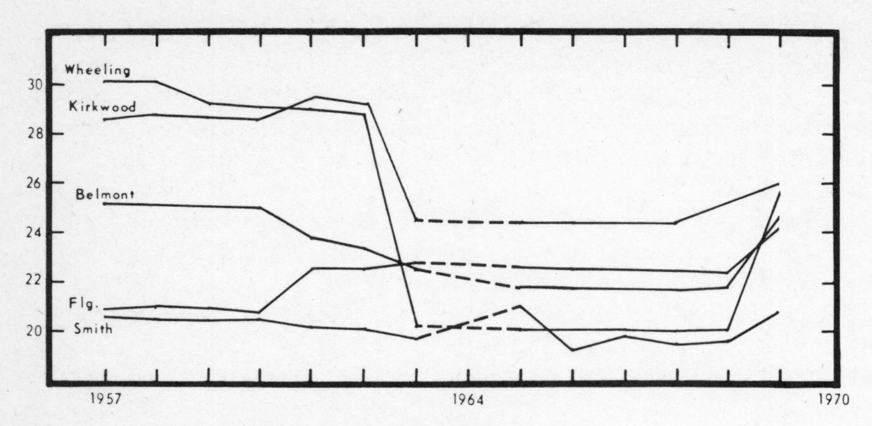

Figure 4 Average assessed value of land per acre

Mrs. Alice Grossniklaus (1968) of the Community Council for Reclamation has documented the changes in land value and the working population in the region. Sixteen percent of Harrison County is stripped and its cropland is heavily silted; the employment among miners has fallen 29% in one decade; the population declined 15% over two decades; and 42% earn less than the poverty level income of $4000 per year. The school tax on stripped land in Garaway, Ohio, is 40¢/acre/year but on nearby farmland is $2.25/acre/year. It is no wonder that the local governments frequently go bankrupt after strip mining.

Timothy Albright (1971) of Case Western Reserve has also obtained economic statistics on Belmont County. Belmont County strip mines produced 14% of all of Ohio's coal production, more than any other county. In fact three-fifths of the county is owned by or leased to coal companies. Landowners there receive from 0.2% to 4% of the value of the coal when they sell their land to the coal companies. When the company merely owns the mineral rights, the landowner receives nothing. In contrast, royalties to owners of leases on oil wells are typically 12%.

Assessed land values in the region dropped rapidly after stripping began. Figure 4 shows that assessments in nearby unstripped Smith Township held steady, but that the values in Wheeling and Kirkwood Townships dropped precipitously when the $6 million "Gem of Egypt" (Giant Earth Mover) arrived in 1962. Figure 5 shows that the value of buildings consistently declined in the stripped townships while values were rising elsewhere.

Figure 6 reveals that the value of buildings on land owned by Consolidation Coal was assessed only one-third of that on other surrounding land—even more so when the coal processing plants are excluded.

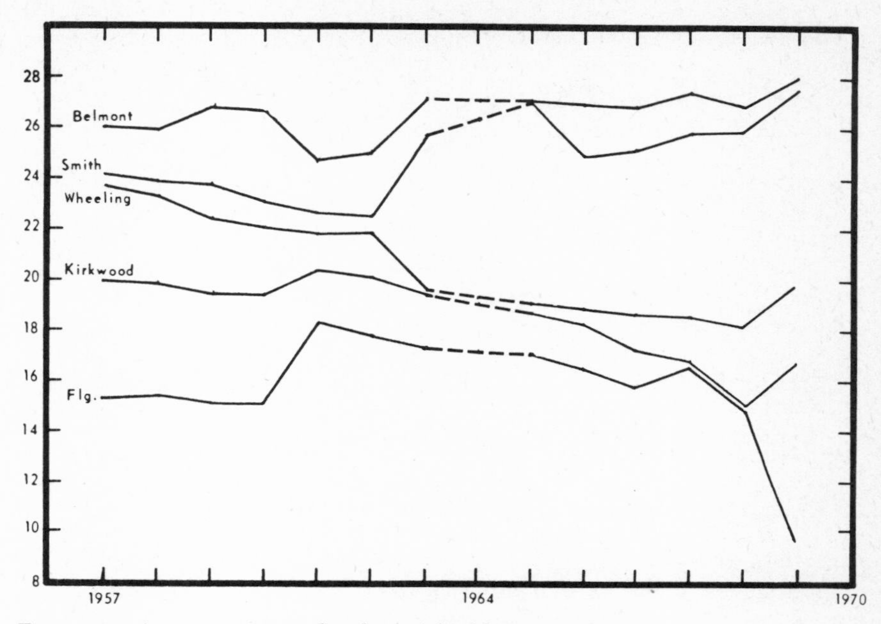

Figure 5 Average assessed value of buildings per acre

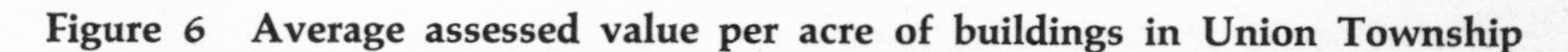

Figure 6 Average assessed value per acre of buildings in Union Township

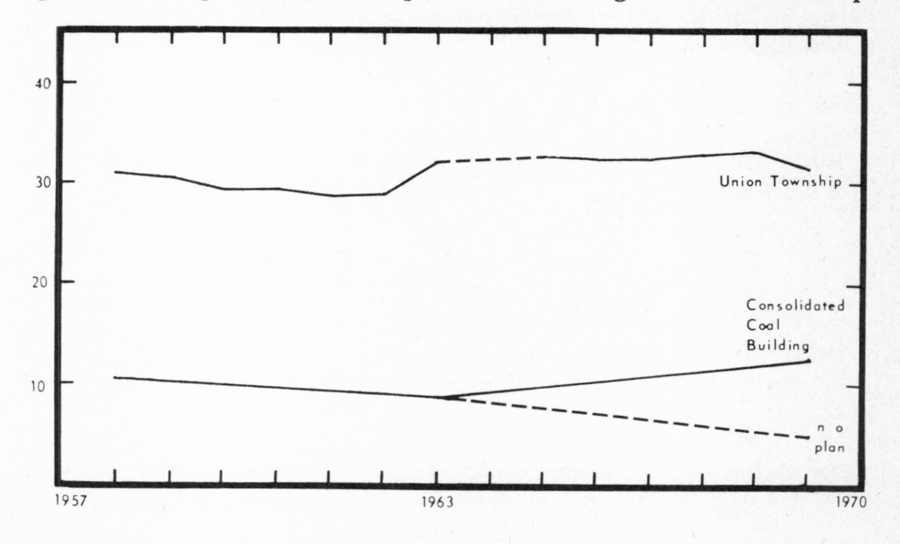

The idea that stripping creates significant local jobs also appears false. Only 561 out of a county population of 80,000 were employed at surface mines (Albright, 1971).

Jack Hill, in another case study (Albright, 1971), concluded that

a) local businesses are not extensively supported by the mining industry,
b) land values and agricultural production have grown at a slower rate than most of the other counties in Ohio,
c) through the sale of coal lands there was no large release of money to Harrison County.

We see, then, that strip mining not only demoralizes a community, but can depress its land values, remove its tax base, and depopulate it of its able-bodied citizens.

An Estimate of Economic Damage from Appalachian Strip Mining

It is, in general, very difficult to determine the costs imposed on the community by pollution and other forms of environmental disturbance. This category of costs is known to economists as "external costs" because unlike the "internal costs" which are borne by the producer they are passed on to others and hence do not influence the production decisions. An attempt to estimate these costs for strip mining was made by Herbert Howard (1971). His estimate is based on five assumptions:

1) That all social costs vanish within a few years after mining. For example, if no reclamation is performed, the damage estimate for acid leached off the mining bench and coal face falls from $6.88 per acre during the first year after mining is completed, to zero after six years.
2) That future costs incurred by mining during a "base year" would be discounted 12% per year after the base year.
3) That a low value should be placed on aesthetic loss in a sparsely populated area. He assigns the value $1.00 per acre to the loss in aesthetic values caused by Appalachian stripping.
4) That landslides cause an average damage of $50 each.
5) That the only losses suffered by nearby property owners are extraordinary events such as boulders rolling down the hillsides.

Each of these assumptions minimizes the estimate of external costs, especially the first two. Assumption of a 12% annual discount rate for instance, means that a community cost of $1000 related to conditions ten years after the mining operation is evaluated as a cost of only $278.50. Also, Howard did not consider the loss of potential timber or potential farmland, nor any flood damage traceable to strip mining. He does not consider the destruction of ecosystems.

On the basis of these minimizing assumptions, Howard calculates the external costs of strip mining as $91.74/acre (4.8¢/ton) in 1961 and $61.10/acre (2.7¢/ton) in 1967. Most of the damage is due to sediment washed from the spoil banks. Howard claims that Kentucky's rather minimal reclamation laws have caused a one-third reduction in external costs per acre over those years, although the total damage has increased (because of accelerating coal production).

On the basis of such social cost minimizing estimation methods, coal operators have often argued that any substantial expenditures for reclamation are not justified.

Reclamation of Mined Land

The "reclamation" of strip mines means the elimination of off-site environmental damage and the return of the mined land itself to productivity—biologic, economic, or recreational. It does not mean a restoration to its previous ecological state. Former Secretary of the Interior Stewart Udall (1968) made this clear when he testified:

> Let me make the record clear about our use of the word 'reclamation.' In the context of surface mining we do not consider reclamation to mean a restoration of the land to its original condition. Often this would not be as desirable as some alternative land condition. Rather, we use reclamation to mean that activity which avoids or corrects damage to the lands and waters of the vicinity and leaves the area in a usable condition. In some instances it can be more productive than it was originally.

Reclamation efforts range from a mere aerial reseeding of spoil banks all the way to nourishment of lush growth in the original topsoil. The main categories of reclamation work are, in order of increasing complexity, *backfilling* the site to create a gentler topography, prevention or treatment of acid drainage, reseeding (or rarely, replanting), and a replacement of the soil strata back in their original order.

The purposes of regrading a mine site are to eliminate trenches that cause standing pools of highly acid water to form, to cover the coal seam and as much pyritic material as possible (the iron pyrites are the source of acid water), to minimize the seepage of water through loose spoil (which contaminates groundwater), and to smooth the site into a gentle grade so that stormflow and erosion are checked.

The Bureau of Mines has experimentally determined the costs of rudimentary reclamation by the method of backfilling for an old strip mine in north central Pennsylvania. Figure 7 is a cross section of the site before stripping, immediately after stripping, and after partial backfilling done just after mining. The original 14° slope has been reduced to a spoil trench through which water was percolating; the highwalls stood 60 to 80 feet high. Figures 8 and 9 illustrate the five procedures followed; the minimal aim in each case was to grade the spoil bank away from the highwall so that water would not collect amidst the toxic spoil.

In method A the mining site is graded practically flat, with a nominal slope of 3°. The height of the highwall is reduced by half.

Figure 7 Typical strip pit in north-central Pennsylvania
(A) Before Stripping, (B) After Stripping, and (C) After Primary Backfilling and Replanting.

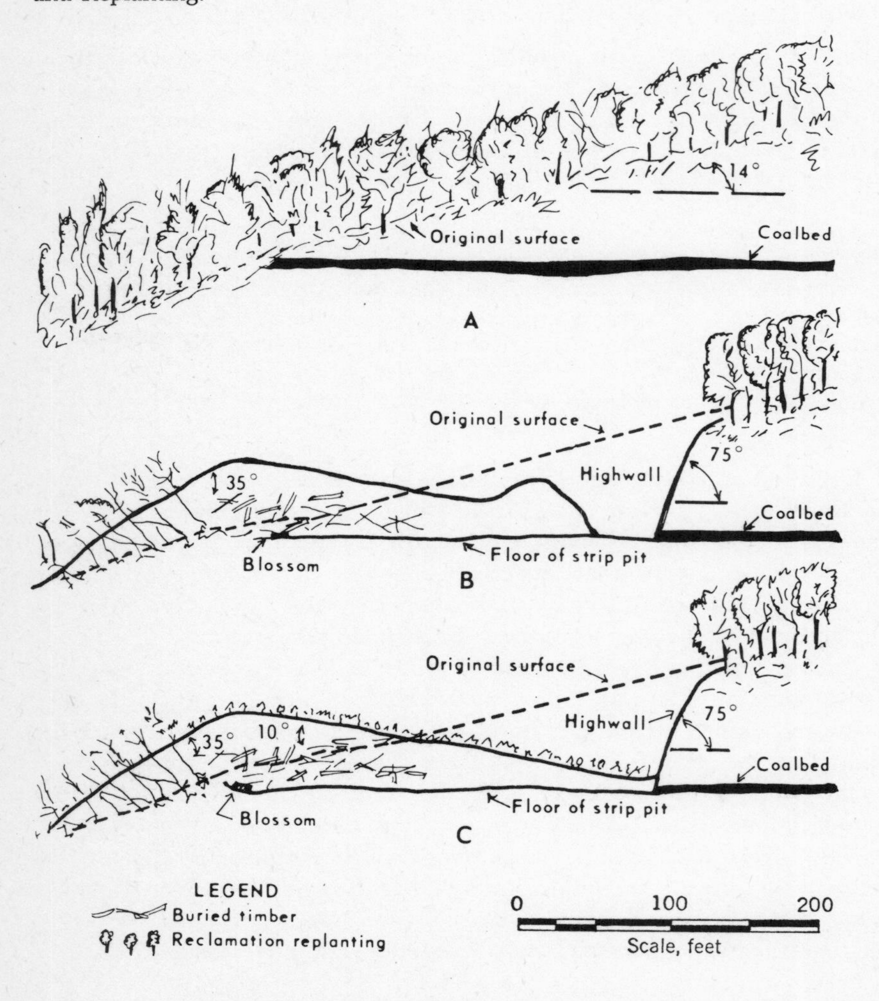

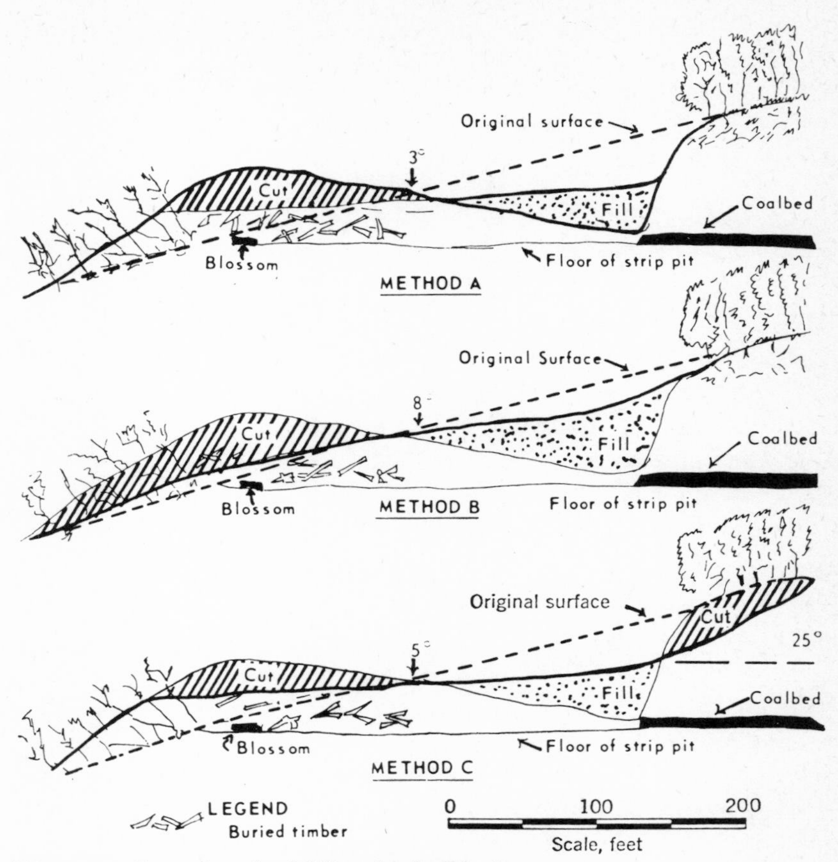

Figure 8 Secondary backfilling by bulldozing

Method B is a restoration of the contour to its original grade, with elimination of the highwall. It involves cutting many buried timbers and shoveling earth up the hill.

Method C is similar to the first except that earth at the top of the highwall is shoveled into the pit. The slope behind the former wall becomes 25°, but the high 35° incline of the outermost spoil bank remains.

The remaining two procedures involve dynamiting the highwall. Method D changes the wall into a 25° slope, and Method E into a 45° angle.

The costs are summarized in Table 3. The cost of Method B, restoration of contour to original grade, could probably be cut by two-thirds if the reclamation were preplanned and performed immediately after mining. The costs of all methods are higher due to timbers buried in the spoil and to soil compaction in the intervening years.

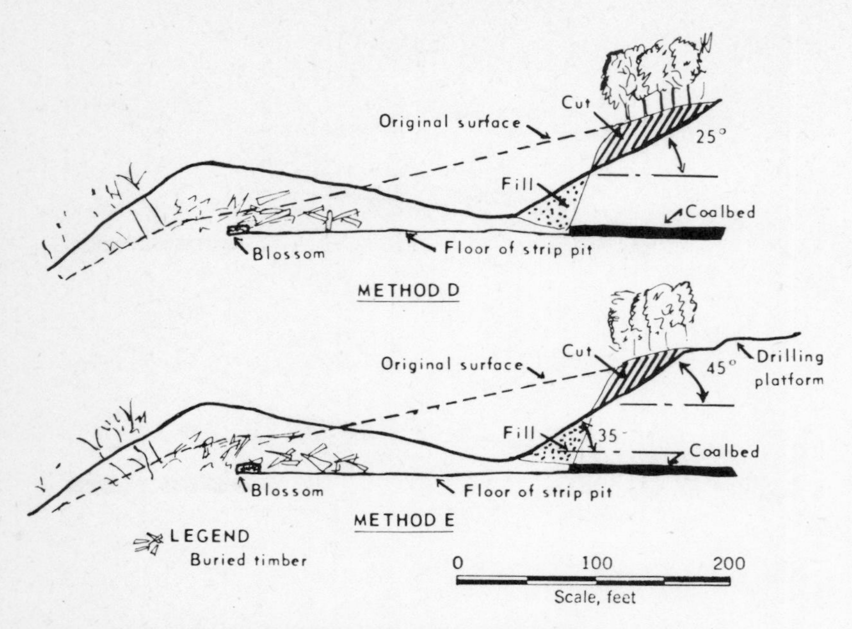

Figure 9 Secondary backfilling by blasting (methods D and E)

A more recent report (Brock and Brooks, 1968) placed the following cost estimate on the above reclamation work: Method A, $250/acre (6¢/ton of coal); Method B pre-planned, $400/acre (10¢/ton); Method E, $460/acre (11¢/ton).

Regrading alone will not prevent acid drainage from strip mines. A good vegetative cover will suffice at some sites, but at others the contaminated runoff can only be checked by dams. At still others, culverts can channel away stormwater before seepage through the spoil renders it toxic.

The lethal acidity of strip-mine spoil can be counteracted with limestone or, ironically, fly ash from the burning of coal (poetic justice?). The Bureau of Mines (Adams et al., 1971) experimented with fly-ash as an aid in reclaiming persistently barren mine sites. The report that 800 tons/acre of fly ash can raise the pH of the soil from 2.9 to 8.2, when mixed into the ground to a depth up to a foot. Since a desirable pH is around 6, this represents an overshoot, but the alkaline component of fly ash tends to leach out over the years. After heavy fertilization (1000 lbs per acre initially, and further application during the season) and watering, the site would *sometimes* yield a stand of grasses and legumes. Trees, and grass on a number of plots, failed to survive.

Table 3 Summary of Costs of Five Backfilling Methods

Item	Total Hours	Total Feet	Hr/ft	Cost/hr	Cost/ft
Method A: Bulldozers	213	570	0.37	$14.00	$ 5.18
Method B:					
Labor	690	370	1.86	2.90	5.40
Bulldozers	244	370	.66	14.00	9.24
Powersaws	404	370	1.09	1.00	1.09
Total	—	—	—	—	15.73
Method C:					
Labor	540	630	.86	2.90	2.49
Bulldozers	398	630	.63	14.00	8.82
Powersaws	246	630	.39	1.00	.39
Total	—	—	—	—	11.70
Method D:					
Labor	1,040	850	1.22	2.90	3.54
Bulldozers	109	850	.13	14.00	1.82
Powersaws	500	850	.59	1.00	.59
Drill	101	850	.12	22.00	2.64
Explosive	58,320[a]	850	68.61[b]	.08[c]	5.49
Total	—	—	—	—	14.08
Method E:					
Labor	639	660	.97	2.90	2.81
Bulldozers	122	660	.18	14.00	2.52
Powersaws	232	660	.35	1.00	.35
Drill	75	660	.11	22.00	2.42
Explosive	6,100[a]	660	9.24[b]	.08[c]	.74
Total	—	—	%	—	8.84

[a]Pounds. [b]Pounds/foot. [c]Dollars/pound.

Although a growth of vegetation admits a strip-mine site as a productive component of the biosphere, and mutes the visual shock of barren, broken land, the foremost purpose of re-planting as a reclamation measure is stabilization of the soil. A sod of tough, quick-growing grass prevents erosion more effectively than a stand of trees. Species of grass found most suitable were Kentucky 31 fescue (*Festuca arundinacea Schrebe*), timothy, (*Phleum pratense*), perennial rye (*Lolium perenne*), red top (*Agrostis alba*), and orchard (*Dactylis glomertas*). Legumes make a good choice because bacteria in their root nodules can fix atmospheric nitrogen into biochemically active ammonia; suitable types are crown vetch (*Vicia*) and birdsfoot trefoil (*Lotus corniculatus*).

The Department of the Interior (1967) estimated that the cost of replanting the 55% of all strip-mined land that needs such attention would average $232 an acre. This is 3.2¢ per ton of coal mined from those areas. A company called Precision Aerial Reclamation (1968) claims that it can reseed and fertilize mined land from airplanes for $35 per acre. The Bureau of Mines (1970) has reclaimed an area in Pennsylvania for use as a state park for $814 per acre (11¢ per ton of coal) where grasses were sown, and for $1448 per acre (20¢ per ton) where pine and spruce trees were sown. The operations also included backfilling to the original contour, burying 45-50 feet highwalls, and covering acid-forming material.

Hill (1969) has presented cost estimates in the range $100 to $430 per acre for revegetating 710 acres of land as in Table 4.

Table 4 Cost of Revegetation[a]

	Dollars Per Acre		
	Maximum	Minimum	Average
Conventional Grass[b]	462.08	71.13	165.22
Hydroseeding Only[c]	370.72	238.60	323.43
Trees Only[d]	194.76	55.02	106.73
Hydroseeding and Trees[e]	537.06	362.95	429.40
Conventional Grass and Trees[f]	380.53	204.47	230.56

[a]Cost includes labor, equipment, materials, and overhead (Cost distributed on direct labor basis.)

[b]Fertilizer (0.5 ton/acre of 10-10-10), lime (2-4 tons/acre) applied from truck, grass planted by seeder box

[c]Lime (2-4 tons/acre) spread from truck or from farm type fertilizer spreader, hydraulic application of grass seed, fertilizer (0.5 ton/acre of 10-10-10) and organic mulch (1 ton/acre)

[d]Hand planted (900-1000/acre)

[e]Hydroseeding plus hand planted trees (900-1000/acre)

[f]Conventional grass as in b, plus hand planted trees (900-1000/acre)

Even if human intervention could establish perfect soil conditions for regrowth, there is little hope of regaining the quality of the forest-wildlife community as it was before it was annihilated. A region where life has been drastically disturbed, or where it has yet to become established, undergoes a gradual process called *succession*. A few hardy small plants such as lichens or grasses gain the first foothold and prepare the land for a series of plant communities, each a little larger and more complex than the preceding ones. Under many circumstances, a *climax* ensues, characterized by a community of species that dominates the landscape and which can persist indefinitely. An ecosystem undergoing succession typically has high productivity: much energy and inorganic material is transformed into standing biomass instead of being recycled as in a mature climax state. For strip-mine reclamation this means that newly-planted areas must either be heavily fertilized or sown with a leguminous cover, as spoil banks frequently are devoid of nitrogen and low in phosphorus and potassium. But more importantly, the slow pace of succession implies that we will not enjoy mature well-forested reclamation sites in this century, at least. The length of time between the disruptive event and the attainment of a stable temperate forest climax is around 150 years on land whose humus is left intact and where both microclimate and soil type are favorable to rapid growth. The process known as secondary succession occurs on clear-cut forested areas and abandoned cropland. Primary succession on sand dunes and, presumably, on bare rock requires about one thousand years (Odum, 1971). Since the American practice is to bury the original topsoil under tons of overburden and leave in its place acid earth devoid of humus, the reversion time on stripped land will likely lie between 150-1000 years.[2] And since strip-mining is a 20th century creation, no mined land has attained a climax.

This is particularly unfortunate in certain parts of the country. The productive central Appalachian coal fields, centered in eastern Kentucky, also happen to sustain the most advanced temperate forest ecosystem on this side of the world. The mountainous region, blessed with 50 inches of rainfall annually and excellent drainage, sustains over 2000 types of seed-bearing plants and 1000 species of trees, more than anywhere else in the Northern Hemisphere except perhaps Eastern China.

The North Germans practice the least environmentally destructive mode of strip mining. They scrape off the topsoil and store it, and do

[2]Where the topsoil is carefully replaced, the reversion time would lie close to 150 years; if not, the time will be much longer. United States mining companies do not ordinarily replace the topsoil on mine sites as the Europeans do.

the same with the subsoil, before removing the coal. After the coal has been extracted they replace the strata and soil in its original order; plants then grow readily in the topsoil (Gärtner, 1968). The fertile soil is returned to the mine surface in prepared polders after mixing with water at a 1:1 ratio. After the soil dries it is ready for cultivation. Windbreaks keep out the severe frost during the initial critical period of regrowth. Much of the area is returned to agriculture, and in the rougher areas, "The slopes and banks of the dumps are being afforested and the forest which results is made available to the local population for use as a place for relaxation and recreation." (Gärtner, 1968).

Admittedly the German environment responds readily to this type of reclamation. The depth of the loess topsoil[3] is measured in meters rather than inches, and the deep sandy subsoil readily harbors tree roots. The land prepared for recultivation should not have a slope greater than 1.5%, a requirement that the Rhineland meets. As the seams of lignite can be hundreds of feet thick, the company affords $8000-$10,000 per acre for all reclamation; their average expense for replanting alone is $405 per acre (Dobson, pers. comm.). This can be economical because of the great thickness of the lignite.

Let us compare the possible with what is actually done in American strip mining. Most states with coal resources have reclamation laws on the books.[4] They require operators to purchase licenses costing a few hundred dollars, and to post a bond averaging $2100 plus $350 per acre mined, which is refunded after satisfactory reclamation is performed. The figure $350 lies on the low side of reclamation costs, so it is frequently profitable for an operator to abandon the site unreclaimed and forfeit the bond. Only four states (Kentucky, Ohio, Tennessee, West Virginia) have financial penalties (averaging $2300 per operator per day) for flouting the strip-mining laws. Most states specify that some grading be done, but only Pennsylvania and Kentucky require that the mine be backfilled to the original contour. About half of the states with laws call for some reseeding to be done, but rarely is the success of regrowth a factor in refunding the bond. Only West Virginia asks for some measure of successful revegetation—it has set a minimum of 600 plants (of any kind, I suppose) per acre or a 70% cover with grasses or legumes.

A probable result of the lack of criteria for regrowth is the fashionable use of "quickie" reseeding methods such as aerial seeding and the dispersal of seeds from water hoses. The Department of the Interior

[3]Loess is wind-blown soil.

[4]State reclamation laws are summarized in the Bureau of Mines Survey of Strippable Coal Resources.

reported wide variations in the enforcement of state laws. As of 1966, the eight most active stripping states employed 62 field inspectors, supervisors, and engineers altogether to police 37,300 actively mined acres. The amount of forfeited bonds from 1963 to 1966 ranged from zero in West Virginia, Illinois, and Indiana to $720,000 worth in Pennsylvania.

Howard (1971), in the report mentioned earlier, calculated the approximate amount that mining companies spent on reclamation in Eastern Kentucky from 1962 to 1967. Of $47 spent per disturbed acre in 1967, $16 represents the cost of regrading and $31 the cost of revegetating. Additionally, the firms spent $38 per acre for such enforcement expenses as permit fees, court fines, performance bonds, and land surveys. The costs translate into 2.09¢ per ton for actual reclamation and 1.71¢ per ton for enforcement.

A number of bills dealing with strip-mining are now before Congress. Representative Ken Heckler's bill would prohibit new strip mines as soon as the bill is enacted, and would ban all strip mining after 2 years. As a federal law it could apply only to those mines that deliver coal or buy equipment through interstate commerce. The bill provides for a program to reclaim all strip-mined land before 1980 with the Federal government paying up to 90% of the costs. The same bill also sets up a timetable for establishing environmental standards on underground mines similar to the program espoused in the Clean Air Acts.

The Ninety-second Congress came very close to enacting federal controls over all strip mining, but did not do so when the Senate and the House disagreed on key provisions of a strip-mine control bill during the last days of the session. The stronger House version, which the coal industry adamantly opposed, would have had the Department of the Interior establish federal reclamation standards for new land strip-mined for coal, unless the State established standards agreeable to the Interior Department. The bill also would have permitted the government to prohibit surface mining on slopes greater than 20°. Such a provision, if enacted, might well have prevented any new strip mining in the vast Appalachian coal province.

The Recovery of Past Mined Lands

Let us first devote our attention to the two million or so acres of strip-mined land which has not been reclaimed and which is not likely to recover naturally in the near future. The Department of the Interior advanced two proposals for a national program to reclaim this land deliberately. The first program, called "basic reclamation," is intended to stop the continuing off-site damage caused by unreclaimed stripped

Table 5 A Basic Reclamation Program

	Basic Reclamation		Limited Program	
Treatment	Acres	Cost	Acres	Cost
Planting	1,760,000	$408,470,000	374,000	$126,190,000
Contour Benches	198,000	40,100,000		
Grading			124,000	18,900,000
Area Strippings	186,000	18,100,000		
Drainage Control	567,000	111,100,000	304,000	59,500,000
Diversion Ditches	4700 miles	10,000,000	4700 miles	10,000,000
Slide Damage Repair	860 miles	12,900,000	860 miles	12,900,000
Haul Road Repair	6000 miles	18,000,000	6000 miles	18,000,000
Stream Channel Repair	7000 miles	21,800,000	2500 miles	10,000,000
Pond Stabilization	500 ponds	2,500,000	500 ponds	2,500,000
Highwall Access	7400 miles	5,200,000	3700 miles	2,600,000
Hazardous Conditions	5000 miles	10,100,000	1900 miles	3,800,000
Total		$658,270,000		$264,390,000
Administration, Engineering and Planning @ 15%		$ 98,740,000		$ 39,658,000
Grand Total		$757,010,000		$304,048,000

This is $370 per acre for the basic program. (1966 prices)

land. The principal damage is water pollution from sediments and acid. (See Table 5 for a breakdown of the proposed treatments and cost.) Planting, the most expensive factor in this particular program, is aimed at establishing a quick, protective cover of grasses and legumes to stabilize the spoil rapidly. Certain marginal measures, such as dredging streams, fencing off highwalls, and building one access road for each mile of highwall, may be questioned, but their cost is minor relative to the expense of the whole program. The entire cost is $757 million, or $370 per acre, at 1966 prices. The cost of a more limited program of reclaiming only those areas judged "to be most urgently in need of attention" is $304 million.

Their second proposal calls for a "rehabilitation" of stripped land for specialized uses, ranging from forests and wildlife habitats to cultivation and urban development. Table 6 displays the proposed land uses and the cost of the necessary measures. Notice that the extra expense of upgrading the land to "productive" use is less than the cost of basic reclamation.

Let's hypothesize that these national reclamation programs are done in the next five years and that the expense is charged to the coal industry as payment on their debt to the nation. We accept the figures of the Department of the Interior and assume that the industry will extract 600 million tons of coal per year, as it is now doing. The expense of basic reclamation translates into a 25¢ increase in the cost of one ton, and that of rehabilitation into a 40¢ increase. If one ton of coal sells for $9.00 before freight charges are added, then the price increases will be +2.8% and +4.5% respectively. (Freight charges, though highly variable, are comparable to the price at the mine.) The span of five years would be the shortest practical time in which to reclaim so much land; if the task takes longer than a decade the effect on fuel prices becomes trivial.

Summary of Reclamation

We see that although it is possible to return strip-mined land to productivity, the slow pace of Nature as codified in the findings of the science of ecology precludes quick restoration of the land to its original state. Instead, if mankind dares to scrape minerals out of the earth, he must be content with restoring the contour and reseeding grasses, trees or at best crops. The costs of such reclamation are generally trivial compared to the cost of the coal itself. Extensive regrading costs about 20¢ per ton, good replanting about 5¢ per ton; but coal now sells for about $8 per ton. However, this still leaves us with the question, Is the intensive restructuring of the environment characterized by strip mining,

Table 6 A Rehabilitation Program for Mined Land

Land Use	Percentage of Reclaimed Area	Thousands of Acres	Cost	Necessary Measures
Cropland	2%	43	$ 29,141,000	Grading, building drains, preparing topsoil, fertilizing
Forests	23%	465	167,748,000	Nothing beyond basic reclamation
Wildlife	21%	439	181,489,000	Planting for game food and habitat
Ponds-Lakes	6%	145	147,946,000	New ponds build for $1000/acre
Streams	7%	136	34,738,000	Construct sediment basins, spawning areas, rip-rap
Recreation	11%	217	151,125,000	Only potable water supplies and sewage facilities
Pasture	12%	242	149,937,000	Grades must be 15%. Establish quality grasses and legumes
Rangeland	12%	233	121,477,000	Strike-off grading and spot planting
Occupancy	6%	121	69,540,000	Intensive grading
Cost Breakdown by Nature of Work				
Basic reclamation		$ 658,270,000		
Construction of new impoundments		145,446,000		
Grading		163,536,000		
Planting		42,311,000		
Services		43,578,000		
Total		$1,053,141,000		
Administration, engineering, and planning @ 15%		157,971,000		
Grand Total		$1,211,112,000	$593/acre	

(1966 prices)

and its long-term damage, justified for the transient convenience of some quickly expended electrical energy?

REFERENCES

Aaronson, Terri, 1970. Problems underfoot. In *Environment* 12, no. 9 (Nov.). Aaronson obtained from the Department of the Interior a "working draft" of the report *Environmental effects of underground mining and of mineral processing.*

Adams, L. M., et al., 1971. Bureau of Mines Report of Investigations 7504, Reclamation of acidic coal-mine spoil with fly ash. Apr. The experiments are poorly executed—the control plots appear to be treated differently from the others, unfavorable data tends to be excluded from the conclusions, and some fudging is suggested.

Albright, Timothy A., 1971. The hidden costs of strip mining: a socioeconomic study of Belmont County. Intersession Project, Case Western Reserve University, Cleveland, Jan.

Austin and Borrelli, 1971. Fall offensive against stripping begins in the House, *Environ. Action* Sept. 18.

Averitt, Paul, 1970. U.S. Geological Survey Bulletin 1322.

Biesecker, J. E., and J. R. George, 1966. Stream quality in Appalachia as related to coal mine drainage. U.S. Geological Survey, Circular No. 526, 1965.

Brock, S. M., and D. B. Brooks, 1968. The Myles Job Mine—a study of benefits and costs of surface mining for coal in Northern West Virginia. Research Series 1, Office of Research and Development, Appalachian Center, West Virginia University (Morgantown), Apr.

Bureau of Mines, 1970. Information Circular 8456, Surface Mine Reclamation. Moraine State Park, Pennsylvania.

Dobson, Edward, personal communication. Mr. Dobson is with Friends of the Earth, London, England.

Federal Water Quality Administration, June 1970. Clean water for the 1970's —a status report.

Gärtner, Dr.-Ing. E., 1968. Opencast Mine Garsdorf, to the II International Surface Mining Conference in Minneapolis, Minnesota, September. Dr. Gärtner is with the Rheinische Braunkohlenwerke AG, Cologne, Germany.

Griffith, F. E., et al., 1966. Demonstration and evaluation of five methods of secondary backfilling of strip mine areas. Bureau of Mines Report of Investigations 6772.

Grossniklaus, Alice, 1968. *Testimony in surface mining reclamation.* Hearings before the Committee on Interior and Insular Affairs, U.S. Senate on S. 3132, S. 3126, and S. 217, Apr. 30, May 1 and 2, 1968, p. 283.

Hechler, Ken. Hechler's principal bill on strip-mining is H.R. 4556, the *Environmental Protection and Enhancement Act of 1971* in the Congressional Record, Feb. 18, 1971, p. H805.

Hill, R. D., 1969. *Reclamation and revegetation of 640 acres of surface mines: Elkins, West Virginia.* Paper presented before the International Symposium on Ecology and Revegetation of Drastically Disturbed Areas, Penn. State Univ., Aug.

Howard, Herbert A., 1971. A measurement of the external diseconomies associated with bituminous coal surface mining, Eastern Kentucky, 1962–1967. *Nat. Resources J.* 11 (Jan.): 76.

National Coal Association, 1970. Bituminous coal facts.

Neely, James C., III, 1970. The effects of strip mining on a natural system: A water quality study of Piedmont Lake, Ohio. Intersession Project, Case Western Reserve Univ. Cleveland, Jan.

Odum, E. P., 1971. *Fundamentals of ecology.*

Precision Aerial Reclamation, 1968. *Surface mining reclamation.* Hearings before the Committee on Interior and Insular Affairs, U.S. Senate on S. 3132, S. 3126, and S. 217. Apr. 30, May 1 and 2, 1968.

Udall, Stewart, 1968. *Surface mining reclamation.* Hearings before the Committee on Interior and Insular Affairs, U.S. Senate on S. 3132, S. 3126, and S. 217, Apr. 30, May 1 and 2, 1968, p. 36.

U.S. Department of the Interior, 1967. *Surface mining and our environment.* A special report to the nation.

U.S. Geological Survey, 1970. Professional papers 427-A, 427-B, 427-C.

MORTON MAY, ROBERT LANG, LEANDRO LUJAN,
PETER JACOBY, AND WESLEY THOMPSON

Reclamation of Strip Mine Spoil Banks in Wyoming[1]

Approximately 15,000 surface mines on more than three and a half million acres of land, occur in the United States. Many are scattered throughout states west of the 100th meridian. From the standpoint of plant growth, climatic conditions in these states, with the possible exception of Hawaii, are extreme. Seventy-five percent of the area receives less than 20 inches annual precipitation and 20% of it receives less than 10 inches of precipitation annually. It is estimated that in many areas, losses from sublimation and evaporation leave less than 50% of the measured precipitation for plant growth.

Superimposed on this picture of limited precipitation are seasonal temperature ranges from −60° to 120°F, short frost-free periods, and wide variations in overburden material from surface mines.

Problems of revegetating strip mine overburden piles in western United States differ from those in more humid regions such as Appalachia. Because there has been no research in this area or under these climatic conditions, a cooperative study was initiated with the Kemmerer Coal Company in 1964. The study had the following objectives:

1. To determine adaptability of native or introduced plant specis for revegetating overburden piles;
2. To determine if fertilization, mulching, snow fencing for water accumulation, and/or various mechanical soil treatments would significantly affect vegetation establishment and growth.

[1]This chapter was adapted from May, M., et al., "Reclamation of Strip Mine Spoil Banks in Wyoming," *Research Journal,* 51, Ag. Exp. Sta., University of Wyoming, Laramie, September 1971. The authors express their appreciation for the support of the Kemmerer Coal Company.

Most spoil reclamation research had been conducted in more humid climates than Wyoming's and usually revegetation attempts were with tree species.

THE STUDY AREA

The Kemmerer coal fields are in southwestern Wyoming about four miles southwest of the city of Kemmerer.

The terrain is rolling hills and ridges separated by broad open flats. The strip mines lie on a ridge caused by the Lazeart Syncline and overlook Hilliard Flats which is traversed by many intermittent streams.

The study area is a part of the Northern Desert Shrub region and is dominated by big sagebrush (*Artemisia tridentata*), other shrubs, and several grasses associated with those shrubs. Vegetation on spoil banks, while sparse, is characteristic of early stages of secondary succession in that most species are annuals and introduced. On younger spoil banks, Russian thistle (*Salsola kali*), fireweed summercypress (*Kochia scoparia*), and cheatgrass brome were common. On older spoil banks, bottlebrush squirreltail, foxtail barley (*Hordeum jubatum*), and a few native shrubs have become established.

Annual precipitation in the study area is unpredictable with periodic drought the rule rather than the exception. *Based on a 32-year average for Kemmerer, annual precipitation is 9.42 inches.* Much of the total annual precipitation occurs as snow with an average annual fall of 56.5 inches. Highest precipitation is in spring and early summer with May and June averaging slightly over one inch each. Precipitation during this period occurs mainly as rain.

Precipitation as snow has several peculiarities. It usually occurs after the ground is frozen so any snow which melts is subject to runoff rather than percolation into the soil. Snow also is blown about and distributed in uneven patterns. Many areas catch little snow while others such as gullies and leeward sides of wind obstructions receive large amounts. Snow also is vulnerable to sublimation, a process by which solids pass directly into a gaseous state without being transformed to liquid. It is not uncommon for 60 to 80% of a snowfall to be sublimated, leaving 20 to 40% of the moisture content to be transformed into water, which may or may not penetrate the soil surface. It is not unreasonable to estimate that of the 9.42 inches of annual precipitation, less than five inches are available for plants.

Based on long time records, average annual monthly air temperatures range from a low of 12°F in January and February to a high of 63°F in July. These temperatures, recorded at Kemmerer, can be related to similar climatic conditions at the Kemmerer coal field, approximately four miles to the southwest.

Air and soil temperatures at the study area vary with exposure, topography, and microclimatic level, between different color combinations of the over-burden material.

The study area is unfenced rangeland, and is periodically grazed during spring and early summer by bands of herded sheep. Stock trail over the spoil banks and use water trapped in pits, all of which are accessible. Study plots are unfenced and subject to grazing as would be the entire area if revegetated.

Wildlife also live in the study area, using pit water when necessary. Deer are abundant and according to mining personnel are increasing in numbers.

Spoil Characteristics

Spoil material of the Kemmerer coal fields is derived from the Adeville Formation, local name for part of the Mesa Verde Formation of Upper Cretaceous age. Average formation thickness is about 6000 feet and contains some 30 coal seams ranging in thickness from two to 120 feet.

The spoils of the study area were quite different from soil in surrounding, undisturbed areas. The tremendous disturbance caused by deep cuts and mixing of subsoil strata during mining operations created extreme variability in spoil characteristics within the stripped area. Variation in pH, for example, ranged from 2.2 to 7.3. These extreme values were observed within a distance of three to four feet. The study area as a whole, however, according to Finn's rule (1958), was classified as acidic; only five out of 21 pH readings had pH values greater than 7.0, but no values above 7.3 were recorded in the area (Table 1).

Some spots, having "wet" or "greasy" appearances, were quite high in aluminum content and were extremely low in pH. No vegetation was observed on such spots.

Overburden material contained large amounts of clay. The most frequently found soil textures were clays and clay loams. Spoil materials were high in soluble salts. Calcium and magnesium were quite high, especially when compared with adjacent native soils.

The amount of organic matter, according to soil analyses, was very high, but coal present in spoil material caused analyses to show more organic material than was actually present.

Topography of Spoil Banks

The strata overlying coal seams had been removed by dragline and either transported by truck to a dumping site or stacked by the dragline near the removal site. Differences in methods of deposition re-

Table 1 Average Seedling Number per Square Foot for Each Treatment and Age Group

Treatment	Spoil Age Group: 15 Years		9 Years		3 Years		Total Seedlings for Treatment
	Number of Seedlings	Percent of Highest Number	Number of Seedlings	Percent of Highest Number	Number of Seedlings	Percent of Highest Number	
Jute-mulch	255	100.00	142	55.68	14	5.49	411
Mulch	240	94.11	105	41.17	—	—	345
Jute	221	86.67	93	36.47	13	5.09	327
Snowfence-jute-mulch	138	54.11	85	32.33	33	12.94	256
Snowfence-jute	110	43.13	83	32.54	18	7.05	211
Snowfence	64	25.09	55	21.56	11	4.31	130
Control	35	13.72	15	5.88	1	0.39	51
Total seedlings for age group	1063		578		90		

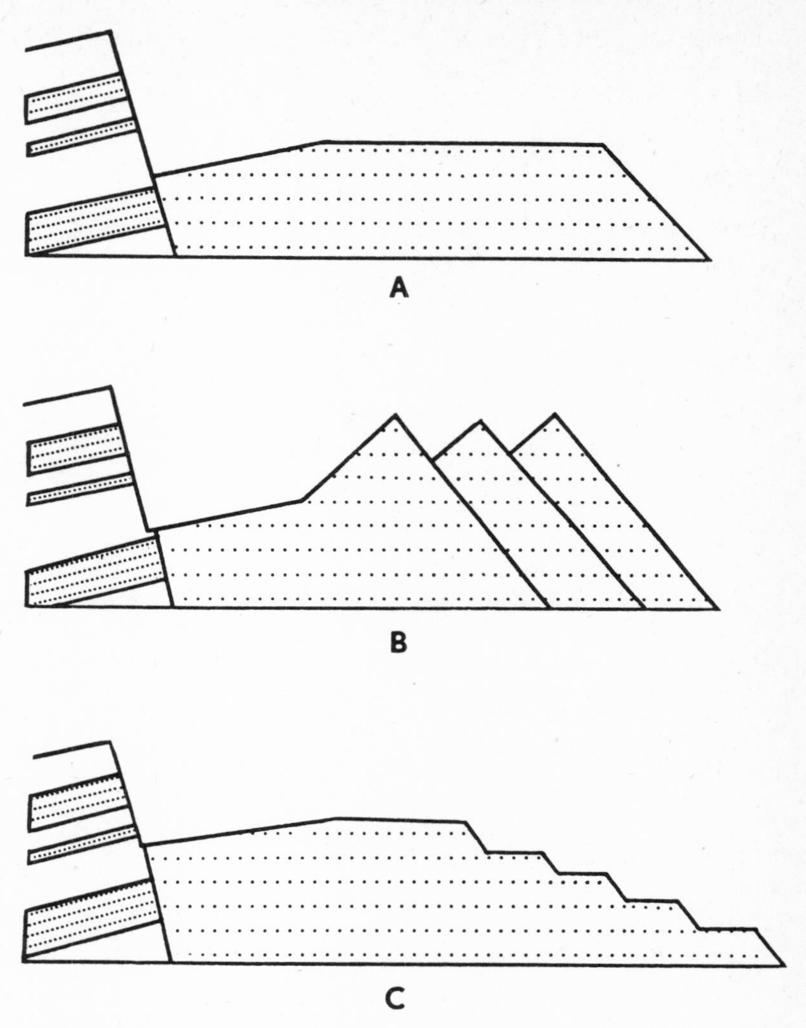

Figure 1 Profiles of spoil bank types (A) truck and dozer type; (B) dragline type; (C) terrace type.

sulted in spoil piles of varying profiles. Figure 1 shows the type of bank with a flat top and solitary long slope (A) and the conical shaped spoil banks formed by the dragline deposition (B). The former is formed by trucking the spoil from another site, dumping it near the edge, and having bulldozers level the top to enable further deposition.

From a revegetation viewpoint, several disadvantages lie in these two types of spoil banks. Erosion is very pronounced and the possibility of seeding and mulching with mechanical means is nil. Hand seeding and mulching is costly and the manipulation of mulching material on the slopes is difficult and often hazardous. Irrigation on such spoil banks is impractical, if not impossible.

The third type of spoil bank (C), without extremely steep slopes, would probably present the best opportunity for successful revegetation. It would enable mechanical seeding and mulching and also would offer good possibilities for irrigation. To conform old spoil banks to this profile would be expensive and would uncover unoxidized material that would slow revegetation.

Topsoil, although not essential for plant establishment, may shorten the time lag in plant succession. In natural revegatation on a 15-year old spoil bank, plant growth from voluntary species has nearly stabilized the bank against erosion.

Among the topographic modifications of the study area from surface mining are numerous pits. Although reclamation plans call for filling these pits with spoil, many have beneficial value as natural reservoirs. Water from ground seepage and precipitation becomes trapped and serves as a source of water for livestock, wildlife, and irrigation. Most water in the Kemmerer coal field area has been tested and declared suitable for livestock and agricultural uses. Fish have been transplanted in selected pits, and salamanders are found in many ponds.

CHARACTER OF DIFFERENT AGED SPOIL BANKS

This study observed three different aged spoil banks in an attempt to recognize and describe definite stages of natural plant succession. The three age groups corresponded with those on which treatments were tested.

The youngest spoil was of a very loose material, highly susceptible to erosion, and formed a thin, hard crust on the surface when undisturbed. Vegetation, usually sparse, was similar in composition on slopes from top to bottom, and there was no apparent influence caused by directional exposure. Species invasion from surrounding areas onto the foot of the spoil bank slopes was discouraged by the constant downmovement of the spoil which covers the volunteering species.

The primary pioneer species on young spoil was Russian thistle. It made up 98% of the vegetation and contributed from 15 to 20% ground cover. This species contributed about 95% vegetation observed on spoil banks less than four years old. Other species included Nuttall saltbush, thickspike wheatgrass (*Agropyron dasystachyum*), and bottlebrush squirreltail.

The middle-aged spoil studied was eight years old. More stable on level areas than young spoil banks, the material still was loose and highly subject to erosion on slopes. Plants were clustered where snow accumulation was evident, suggesting moisture availability or seed accumulation by wind. *Total vegetative cover was estimated at slightly over 15%.*

Russian thistle was still the most abundant species: constituting 95% of samples. Bottlebrush squirreltail, cheatgrass brome, Canada thistle (*Cirsium arvense*), and basin wildrye were in lesser abundance and were also found on surrounding areas not affected by strip mining.

Cheatgrass brome occurred in all samples as a highly aggregated species. The mulch it produced seemed to provide suitable habitat for new plant establishment.

The oldest spoil bank in the study area was 15 years old. Like the mid-age spoil, the 15-year-old bank was stable on the top and level areas but, for the most part, was unvegetated on erodable slopes. *Total cover was estimated to be between 30 and 35% with more plant species than the mid-age bank.*

Some portions of the spoil bank were stabilized by cheatgrass, and in these areas scattered plants from the surrounding communities had become established.

In one particular area on the spoil bank, topsoil removed from another area had been deposited. This area, defined by reddish sandstone and clay soils, supported far more vegetation than nearby overburden. Species found in this area included big sagebrush, bottlebrush squirreltail, foxtail barley, thickspike wheatgrass, and cheatgrass.

Vegetation on the spoil bank top showed evidence that wind dispersal was the main influence in plant distribution. Ninety percent of the samples were bottlebrush squirreltail, with foxtail barley and Russian thistle next in abundance. On margins between the spoil bank and native range, species such as rubber rabbitbrush (*Chrysothamnus nauseosus*), basin wildrye, big bluegrass (*Poa ampla*), silver sagebrush (*Artemisia cana*), thickspike wheatgrass, slender wheatgrass (*Agropyron trachycaulum*), Indian ricegrass (*Oryzopsis hymenoides*), and kochia (*Kochia americana*) appeared.

THE RESEARCH

Three types of research were conducted on the Kemmerer coal spoil banks. These types of research were with (1) trees using various species, fertilizer, and irrigation, (2) grass seeds using four species and various means of holding moisture, and (3) transplanting sod chunks and sprigs of two rhizomatous species with different means of holding moisture.

Tree Planting Research

Four sites were selected on two 14-year-old spoil piles. Two were on the flat tops of spoils and two on slopes with different exposures. At each site, four plots were established, each receiving a different treatment: 66 pounds of nitrogen (ammonium nitrate) per acre; or 33 pounds of nitrogen per acre, irrigation, and control. Five tree and shrub species were planted; caragana, sumac, Russian olive, Siberian elm, and red cedar. Each plot consisted of 10 trees of each species, or 200 trees per site.

Analyses of the spoil material and adjacent undisturbed native rangeland were conducted. Soil moisture records were taken during the year to investigate the effect of moisture on survival and growth of the planted species.

A greenhouse study was conducted to determine fertility of the overburden.

A basic survey of vegetation on both the overburden material (14 years old) and adjacent undisturbed native vegetation was undertaken to compare the percentage cover and species composition of these two areas.

Climatic conditions were studied using data from the Kemmerer weather station about four miles from the study area. The frost-free period in this area is approximately 130 days, and the mean annual precipitation is 9.12 inches. Precipitation in 1965, the first year of the study, was 13.76 inches.

Conclusions

1. Russian olive trees responded the best to the spoils environment, based on survival percentage and growth, of all species tested. This was regardless of location. Second was caragana on flat tops and third best was Siberian elm on the northeastern and eastern slopes. Red cedar and sumac survival was very poor at all locations.

2. The most favorable sites, as determined by percentage survival and growth of all species, were the northeastern and eastern upper slopes. Survival records of 80% and higher were recorded on these locations, one year after the trees were planted. Southern and southeastern lower slopes were the most unfavorable sites. The flat tops of the banks were favorable sites initially, except where salts accu-

mulated or on localized acidic spots. Mortality rates were high on the flat tops after the second year of study.

3. Watering the trees during mid and late summer greatly increased survival percentage growth and vigor. Survival differences as high as 50% were observed between trees on irrigated plots and these which were not irrigated, fertilized, or received no treatment (control).

4. The application of fertilizer (ammonium nitrate) did not significantly affect tree survival or growth.

5. Strong westerly winds caused some tree mortality by breaking and causing faster drying of the overburden material, especially on the flat tops of the spoil piles.

6. Extremely acidic conditions existed in localized spots. The pH was very low and aluminum content high and no vegetation was observed growing on these acid spots.

7. Accumulation of soluble salts, due to poor drainage, was observed frequently on the flat tops of the spoils. These conditions drastically reduced vegetative cover and growth. Overburden material averaged about 13 times as much soluble salts as adjacent undisturbed soils.

8. The overburden material contained a much higher proportion of clay than the adjacent, native soils. Thus, spoil banks were readily eroded and poor drainage conditions usually developed on flat locations. In addition when the soil dried, it formed a very hard surface layer which the young seedlings penetrated with great difficulty.

Seeding Research Grass

A study area composed of coal strip-mined overburden was selected for revegetation studies. Initial goals were to determine grass and forb species best adapted to conditions existing on spoil banks and to simultaneously test mulching and irrigation treatments to determine their effect on stand.

Grass species tested included crested wheatgrass, intermediate wheatgrass, Russian wildrye, smooth bromegrass, big bluegrass, Indian ricegrass, bottlebrush squirreltail, thickspike wheatgrass and basin wildrye. Alsike clover was used on an irrigation treatment and a number of native shrubs and forbs were tested on various sites.

Treatments included jute net, barley straw, mulch, snowfence, irrigation, and combinations of these. Treatments were nonreplicated on overburden piles of three ages: three, nine and fifteen years old. It was attempted to determine the best age of a spoil bank on which to initiate seeding. The degree of erosion stabilization and number of established plants formed the basis for evaluation. Results are displayed in Table 1 and Figure 2.

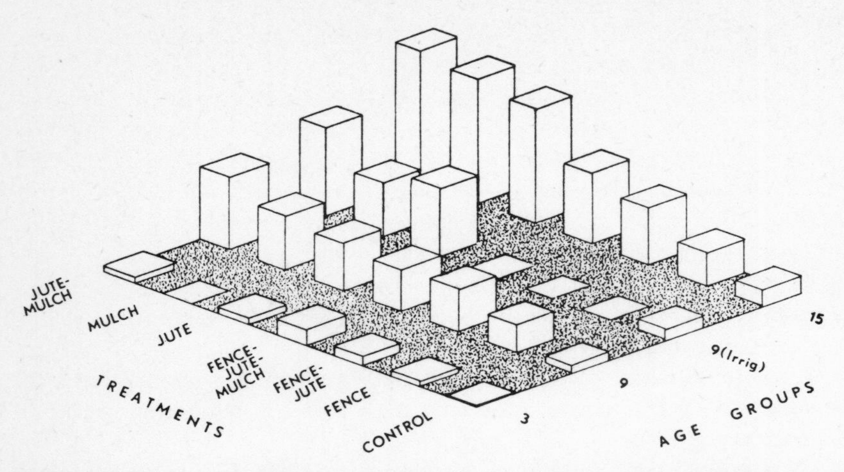

Figure 2 Relative abundance of grass seedlings for different treatments on spoil banks of three ages.

Conclusions

1. Available moisture was a principal limiting factor in plant establishment but this could be supplemented by snowfences and irrigation from nearby permanent ponds. Snowfence was effective only when placed on the leeward side of large, open, level areas.

2. Mulch, necessary for good seedling establishment, required some means of holding it in place. Annual plants grown for a nurse crop served both as mulch and for erosion control but they in turn depended on ample precipitation for optimum growth.

3. Jute netting served as a means of erosion control and as a partial mulch. Stabilization of erosion was a pre-requisite for successful revegetation on slopes.

4. Seedling survival increased with length of weathering period of the spoil pile. Some sites required two years following planting to establish a good perennial grass stand. Annual plants responded more quickly.

5. With irrigation, alsike clover became well established on a spoil bank nine years old.

6. For an economically sound revegetation program, spoil banks should be accessible by machinery for seeding and mulching.

7. Some spoil bank tops revegetated voluntarily within fifteen years. Slopes did not revegetate until stabilized from erosion. Time required for natural revegetation depended entirely on climactic, edaphic, and biotic factors present at the individual sites.

Economic Considerations. Cost considerations are extremely important in reclamation projects and must be considered along with potential

Table 2 Comparative Individual Treatment Costs per acre for Seeding, Material and Labor

Treatment	Cost in Dollars			
	Seed Cost	Material	Man Hours @ $1.25/hr.	Total (per acre)
Jute-mulch	15	1,488.30	91.25	1,594.55
Mulch	15	181.50	91.25	287.75
Jute netting	15	1,306.80	68.43	1,390.23
Snowfence-jute netting Mulch	15	1,628.30	199.68	1,842.98
Fence-jute netting	15	1,446.80	108.43	1,570.23
Snowfence	15	140.00	40.00	195.00
Control	15	0.00	2.50	17.50

results when selecting revegetation treatments. Treatments giving similar results may vary greatly in costs.

Table 2 gives the approximate costs of material used in test plots at the Kemmerer research site. The costs are based on retail costs of material, purchased in small lots for small scale tests. Material in larger quantities would probably result in lower costs per unit.

Grass Transplanting Research

Four study sites on different aged coal strip mine overburden were chosen for revegetation studies to test sprigging and sodding techniques for establishing western wheatgrass and inland saltgrass on different aged spoil piles.

Methods used were planting: at different times of year, by sprigging and sodding, on different aged overburden piles, on tops and on east-facing slopes of piles, and behind and not behind snowfences.

Conclusions

1. Both western wheatgrass and inland saltgrass adapted to vegetative establishment on overburden piles. Western wheatgrass seemed to be the better adapted of the two although inland saltgrass spread more vigorously on favorable sites. Also, inland saltgrass was better adapted to planting during dry weather.

2. The best time of year to plant was in spring. Early fall planting proved least successful. Late fall planting was not tested.

3. Sodding produced far better results than did sprigging. Roots within sod clumps stayed moist and were protected by the surrounding

soil whereas in sprigging, roots were damaged and moisture was lost from plants being prepared for planting.

4. Age of overburden material did not affect planting success. The younger piles showed higher survival rates than did older piles which was the reverse of results obtained from other studies conducted on this overburden material.

5. Planting on top of overburden piles as opposed to the east facing slopes gave better results because infiltration of precipitation and soil moisture were greater on the tops.

6. Planting behind snowfences resulted in slightly better survival because of early spring snow melt behind the fences and the wind break provided by the snowfence. It is doubtful if the slight advantages caused by the snowfence was worth the added cost.

The most limiting factors influencing vegetative establishment was the amount of precipitation received just before, during, and immediately after planting and the amount of moisture in the spoil material at planting time.

GENERAL CONCLUSIONS

1. Ample moisture at planting and during establishment was critical for stand success with seeded grasses, planted trees and grass sod or sprigs. Irrigation and/or the use of snowfences to accumulate extra moisture increased the percentage stand establishment of all types of vegetation.

2. Older spoil piles were better sites for establishing seeded grasses than younger ones. Age of the spoil pile, however, had little effect on establishment of grasses by sodding.

3. Russian olive and caragana were the best adapted tree species tested and the top part of east and northeast facing slopes were the best sites for their establishment.

4. Intermediate and crested wheatgrass appeared to be the best adapted of the cool season grass species seeded. The most satisfactory stands of all species were obtained where mulch with some type of netting to hold it in place was used with the seeding or where the seeding received additional moisture benefits from being on the lee side of a snowfence.

5. Sodded grasses were most effectively established on the flat top of the spoil piles whereas tree species and seeded grasses were more effectively established on northeast and east facing slopes.

6. Nitrogen fertilizer did not significantly affect establishment of either grasses or trees.

COMMENTARY

The preceding assessment of reclamation costs, of course, considers only a part of the large problem of strip mining in the arid areas of the western United States. A more complete documentation of social costs would have to account not only for the cost of revegetation, but also the quality of revegetation as compared to the character of the area before strip mining. In addition, any strip mining operation engenders some degree of erosion by exposing the overburden to the agencies of wind and water. This raises also questions of the contribution of strip mined lands to water quality, as left over mine wastes, eroded material and precipitation runoff may find their way into streams and rivers. Erosion and runoff might actually be exacerbated by the climate of the western states, as long dry periods tend to make soil particularly vulnerable to erosion when precipitation does come, and also leave the soil less able to hold water from running off.

Naturally, the intensity of erosion and runoff problems will vary according to the duration of the interval between the cessation of mining operations and the completion of revegetation, and also according to the quality of revegetation. The social costs or benefits of these environmental effects must be added into the balance sheet in an evaluation of the feasibility of reclamation schemes.

A number of less quantifiable issues also remain to be considered. Where it is appropriate, the opportunity cost of allocating a given area to strip mining, thereby perhaps effectively prohibiting other schemes for utilizing the land until it is sufficiently rehabilitated, and the suitability of various reclamation schemes to the potential future plans for utilizing the land, must also enter into the decisions about reclamation. Finally, there are the aesthetic considerations. Is it enough for a reclamation scheme to merely safeguard the stripped area so that it may be useable in the future, and to minimize environmental effects such as erosion and runoff, or must reclamation include some degree of restoration, so that the area is as pleasing at the end as it was before the mining began?

All of these issues must be raised, and investigated thoroughly before sound decisions can be made about mining the arid western lands.

REFERENCES

Agnew, Allen F., 1966. A quarter to zero. Surface mining and water supplies. *Mining Congr. J.* 52:29–41.

Chapman, A. G., 1967. Effects of spoil grading on tree growth. *Mining Congr. J.* 53, no. 8:93–100.

Czapowskyj, M. M. and W. E. McQuilkin, 1966. Survival and early growth of planted forest trees on strip-mine spoils in the anthractie region. U.S. Forest Service Research paper NE-46, pp. 1–29.

Lorino, L. P. Jr. and G. E. Gatherum, 1965. Relationship of tree survival and yield to coal-spoil characteristics. Iowa State University Agric. and Home Ec. Exp. Sta. Res. Bull. 535, pp. 394–403.

Morgen, Herman Jr. and W. L. Parks, 1967. Reclamation of mined phosphate land. Tenn. Agric. Exp. Sta. Bull. 416, pp. 1–32.

Seidel, Kenneth W., 1961. Seeded black walnut taller than planted walnut on Kansas spoil bank. Central States Forest Exp. Sta. U.S.D.A. Forest Service Station. Note No. 148, p. 1.

Whitt, D. M., 1968. Strip-mined land. How much? What next? *Soil Conserv.* 33, no. 6:123–125.

MAX BLUMER, HOWARD L. SANDERS,
J. FRED GRASSLE, AND GEORGE R. HAMPSON

Oil Spills[1]

During the last few years the public has become increasingly aware of the presence of oil on the sea. We read about the recurring accidents in oil transport and production, such as the disaster of the Torrey Canyon tanker, the oil well blowout at Santa Barbara, and the oil well fires in the Gulf of Mexico. To those visiting our shores the presence of oil on rocks and sand has become an everyday experience; however, few of us realize that these spectacular accidents contribute only a small fraction of the total oil that enters the ocean. In the Torrey Canyon episode of 1967 about 100,000 tons of crude oil were lost. By comparison, routine discharges from tankers and other commercial vessels contribute an estimated three and one-half million tons of petroleum to the ocean every year. In addition, pollution from accidents in port and on the high seas, in exploration and production, in storage, in pipeline breaks, from spent lubricants, from incompletely burned fuels, and from untreated industrial and domestic sewage contribute an equal or larger amount of oil. Thus, it has been estimated that the total oil influx into the ocean is between five and ten million tons per year (Blumer, 1970).

What are the effects of oil on marine organisms and on food that we recover from the sea? Some scientists have said that the oceans in their vastness should be capable of assimilating the entire oil input. This, however, assumes that the oil is evenly distributed through the entire water profile, or water column, of the ocean. Unfortunately, this assumption is not correct. Oil production, transportation, and use are

[1]Reprinted, with permission, from *Environment*, Vol. 13, No. 2, March 1971. The authors acknowledge the continued support of their basic and applied research efforts by the National Science Foundation, the Office of Naval Research, and the Federal Water Quality Administration.

heavily concentrated in the coastal regions, and pollution therefore predominantly affects the surface waters on the continental margins. Ryther (1969) has stated that the open sea is virtually a biological desert. Although the deeper ocean provides some fishing for tuna, bonito, skipjack, and billfish, the coastal waters produce almost the entire shellfish crop and nearly half of the total fish crop. The bulk of the remainder of the fish crop comes from regions of upwelling water, near the continental margins, that occupy only one-tenth of one percent of the total surface area of the seas. These productive waters receive the heaviest influx of oil. They also are most affected by other activities of man, such as dredging, waste disposal, and unintentional dispersal of chemical poisons like insecticides.

Some environmentalists have expressed the belief that major oil spills such as those from the Torrey Canyon and the blowout at Santa Barbara have brought about little biological damage in the ocean (McCaull, 1969). These statements are largely based on statistical measurements of the catch of adult fish. Such statistics are a very insensitive measure of the ecologic damage to wide oceanic regions. Often the migratory history of the fish species studied is unknown. The fish may not have been exposed to the spill itself, or may not have suffered from a depletion of food organisms if their growth occurred in areas remote from the spill. Statistical and observational data on adult fish will not reveal damage to the often much more sensitive juvenile forms or to intermediate members in the marine food chain. The only other studies on effects of oil on marine organisms have concentrated on relatively tolerant organisms which live between the tides at the margins of affected areas. The main impact, however, would be expected in subtidal areas, and that has never been measured quantitatively.

A relatively small oil spill that occurred almost at the doorstep of the Woods Hole Oceanographic Institution at Woods Hole, Massachusetts, gave us the opportunity to study immediate and long-term ecological damage in a region for which we had extensive previous knowledge about the biology and chemistry of native marine organisms (Blumer et al., 1970). On September 16, 1969, an oil barge on the way to a power plant on the Cape Cod Canal came ashore off Fassets Point, West Falmouth, in Buzzards Bay (Fig. 1). Between 650 and 700 tons of No. 2 fuel oil were released into the coastal waters. The oil-contaminated region in Buzzards Bay expanded steadily with time after the accident as the complex interaction of wind, waves, and bottom sediment movement spread oil from polluted to unpolluted areas (Fig. 2). Eight months after the grounding, polluted sea bottom, marshes, and tidal rivers comprised an area many tmies larger than that first affected by the accident. The dispersion was much greater than expected

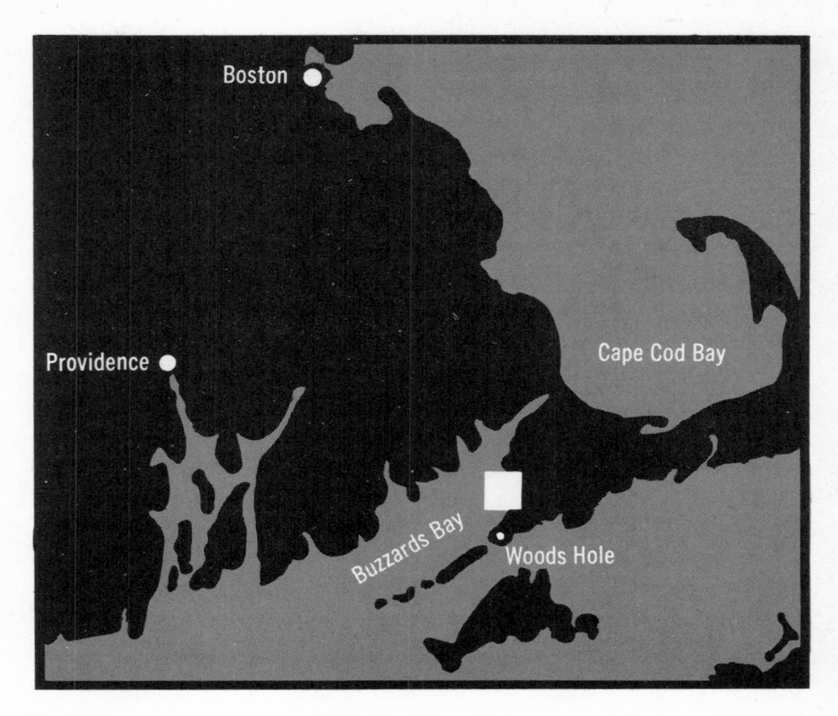

Figure 1 Location of oil spill

on the basis of conventional studies of oil pollution. The situation even forced changes in our research efforts. As we shall explain later, a control point for marine surveys was established beyond the anticipated limit of the spread of oil. Within three weeks, the contamination had spread to the station. Another was established twice as far away. Three months after the accident, that too was polluted. Bottom sediment was contaminated 42 feet beneath the surface, the greatest water depth in that part of Buzzards Bay.

Ecological effects of the spreading blanket of oil beneath the surface were severe. The oil destroyed much of the offshore marine life in the immediate area of the spill during the first few days. As the oil spread out across the bottom of the bay in the following months, it retained its toxicity.

Even by May 1970, eight months after the spill, bacterial degradation (breakdown into simpler substances) of oil was not far advanced in the most polluted regions. More rapid oil deterioration in outlying, less affected areas had been reversed by a new influx of less degraded oil from the more contaminated regions.

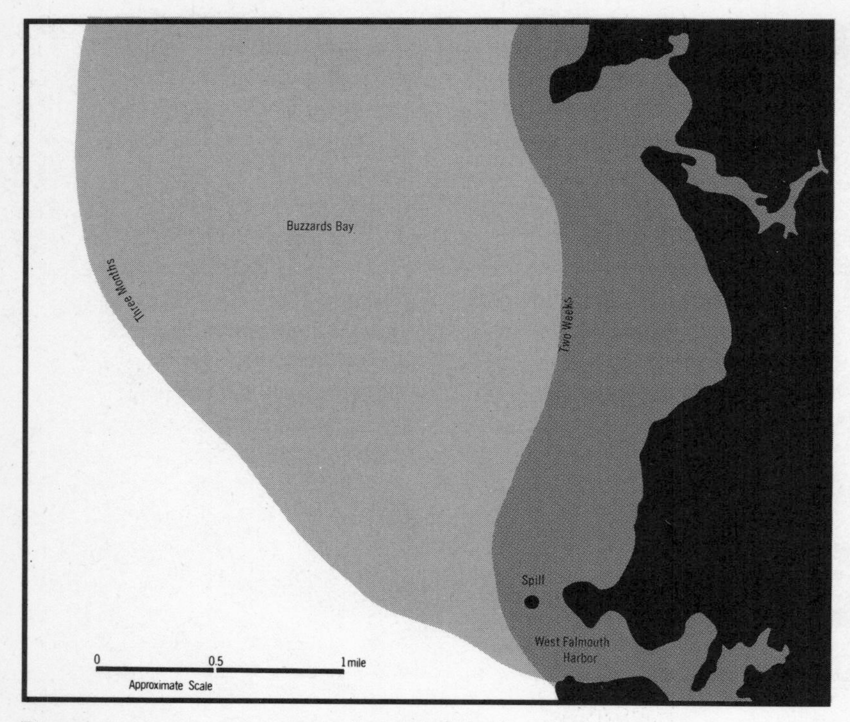

Figure 2 Approximate boundary of the oil contamination at 2 weeks and 3 months after the spill.

The tidal Wild Harbor River still contained an estimated four tons of fuel oil. The contamination had ruled out commercial shellfishing for at least two years. The severe biological damage and the slow rate of biodegradation of the oil suggests that shellfish productivity will be affected for an even longer period. Furthermore, destruction of bottom plants and animals reduced the stability of marshlands and sea bottom. Resulting erosion may have promoted spread of the oil along the sea floor. Inshore, the oil penetrated to a depth of at least one to two feet in marsh sediment.

Nevertheless, compared in magnitude to other catastrophes, this was a relatively small spill; the amount of oil lost in the Torrey Canyon accident was 150 times larger. The interim results of our survey, coupled with research findings of other studies in this laboratory, indicate that crude oil and other petroleum products are a far more dangerous and persistent threat to the marine environment and to human food resources than we would have anticipated. Pollution from a large oil spill is very obvious and visible. It has often been thought that the eventual disappearance of this visible evidence coincides with the disappearance of any biological damage. This, however, is not true. Sensitive analytical techniques can still detect oil in marine organisms and in sediments after the visual evidence has disappeared, and biological studies reveal that this residual oil is still toxic to the marine organisms. Here we shall discuss first the general results of our study, then more deeply the description of the laboratory work involving biology, biochemistry, and chemistry. Our most important findings (Blumer et al., 1970a, b) are as follows.

Crude oil and petroleum products contain many substances that are poisonous to marine life. Some of these cause immediate death; others have a slower effect. Crude oils and oil products differ in their relative composition; therefore the specific toxic effect may vary. Crude oil, in general, is less immediately toxic than some distilled products, but even crude oil that has been weathered (altered by exposure to the weather) at sea for some time still contains many of the acutely toxic hydrocarbons (Blumer et al., 1970a). The more persistent, slowly acting poisons (for example, the carcinogens) are more abundant in crude oil than in some of the lower boiling distillates. These poisons are quite resistant to the environmental weathering.

In spite of low density, oil may mix with water, especially in a turbulent sea during storm conditions. Hydrocarbons may be dispersed through the water column in solution and in the form of droplets, and the compounds may reach the sea bottom, particularly if weighted down by mineral particles. On the sea floor oil persists for long periods and can continue to damage bottom plants and animals. Thus, a single accident may result in long-term, continual pollution of the sea. This is a very important finding since biologists have long agreed that chronic pollution generally has more far-reaching effects than an accident of short duration. Hydrocarbons can be taken up by fish and shellfish. When the oil enters the fat and flesh of the animals, it is isolated from natural degradation processes. It remains essentially constant in amount and chemically intact even after the animals are transplanted into clean water for decontamination. Thus, chemicals from oil that may be poisonous to marine organisms and other animals, including man, may persist in the sea and in biological systems for many months after the spill.

None of the presently available countermeasures can completely eliminate the biological damage of oil spills. The rapid removal of oil by mechanical recovery or by burning appears most promising. The use of sinking agents or detergents, on the other hand, causes the toxic and undegraded oil to spread in the ocean; the biological damage is then greater than if the spill had been left untreated. Reclamation of contaminated organisms, marches, and offshore sediments is virtually impossible, and natural ecological recovery is slow.

With these conclusions in mind we can now turn to our experience with the West Falmouth oil spill. The effect of this relatively small spill was still acute in January 1971, almost a year and one-half after the accident. Officials in the town of Falmouth have estimated that the damage to local shellfish resources, during the first year after the accident, amounted to $118,000. This does not include the damage to other marine species and the expected damage in coming years. In addition to the loss of the oil and the barge and the cleanup expenses (estimated to be $65,000), the owner of the oil paid compensations for the losses of marine fishery resources to the town of Falmouth ($100,000) and to the Commonwealth of Massachusetts ($200,000). The actual ecological damage may far exceed this apparent cost of almost half a million dollars.

BIOLOGICAL AND CHEMICAL ANALYSIS

For our analysis (which is still continuing) bottom samples were carefully taken from the marshes and from the offshore areas. Samples for biological analysis were washed and sieved to recover living or dead organisms. These were preserved, identified, and counted. Results of counts from the affected area were compared with those from control areas that were not polluted by the spill. Some animals can be used as indicators for the presence of pollution, either because of their great sensitivity or because of their great resistance. Thus, certain small shrimplike animals, the amphipods of the family Ampeliscidae, are particularly vulnerable to oil pollution. Wherever the chemical analysis showed the presence of oil, these sensitive crustaceans were dying. On the other hand, the annelid worm, *Capitella capitata,* is highly resistant to oil pollution. Normally, this worm does not occur in large numbers in our area. However, after the accident it was able to benefit from the absence of other organisms which normally prey upon it and reached very high population densities. In the areas of the highest degree of pollution, however, even this worm was killed. *Capitella capitata* is well known, all over the world, as characteristic of areas heavily polluted by a variety of sources.

For chemical analysis, the sediments collected at our biological stations were extracted with a solvent that removed the hydrocarbons. The hydrocarbons were separated from other materials contained in the extracts. They were then analyzed by gas-liquid chromatography. This technique separates hydrocarbon mixtures into individual compounds, according to the boiling point and structure type. To do this, a sample is flash-evaporated in a heated tube. The vapor is swept by a constantly flowing stream of carrier gas into a long tube that is packed with a substance (substrate) that is responsible for the resolution of the mixture into its individual components. Ideally, each vaporized compound emerges from the end of the tube at a definite time and well separated from all other components. A sensitive detector and an amplifier then transmit a signal to a recorder which traces on a moving strip of chart paper a series of peaks (the chromatogram) that correspond to the individual components of the mixture. From the pattern of peaks in the gas chromatogram the chemist can learn much about the composition of the mixture. Each oil may have a characteristic fingerprint pattern by which it can be recognized in the environment for weeks, or even months, after the initial spill. Past and continuing work on the composition of these hydrocarbons that are naturally present in all marine organisms enable us to distinguish easily between the natural hydrocarbons and those contained in the fuel oil. These analyses facilitated our study of the movement of the fuel oil from the West Falmouth oil spill into the bottom sediments and through the marine food chain.

IMMEDIATE KILL

Massive, immediate destruction of marine life occurred off-shore during the first few days after the accident. Affected were a wide range of fish, shellfish, worms, crabs, other crustaceans, and invertebrates. Bottom-living fish and lobsters were killed and washed up on the shores. Trawls made in ten feet of water soon after the spill showed that 95% of the animals recovered were dead and others were dying. The bottom sediments contained many dead snails, clams, and crustaceans. Similarly severe destruction occurred in the tidal rivers and marshes into which the oil had moved under the combined influence of tide and wind. Here again fish, crabs, shellfish, and other invertebrates were killed; in the most heavily polluted regions of the tidal marshes almost no animals survived.

The fuel oil spilled at West Falmouth was a light, transparent oil, very different from the black viscous oil associated with the Torrey Canyon and Santa Barbara episodes. Within days most of the dead animals had decayed and the visual evidence of the oil had almost disappeared. Casual observers were led to report to the press that the area looked as beautiful as ever. Had we discontinued our study after the visual evidence of the oil had disappeared, we might have been led to similar interpretations. From that point on, only continued, careful biological and chemical analysis revealed the extent of continuing damage.

PERSISTENCE OF POLLUTION

Quite recently a leading British expert on treatment of oil spills remarked that "white products, petrol, kerosene, light diesel fuel, and so forth, can be expected to be self-cleaning. In other words, given sufficient time they will evaporate and leave little or no objectionable residue." (Smith, 1970). Our experience shows how dangerously misleading such statements are. Chemical analyses of the oil recovered from the sediments and from the bodies of the surviving animals showed the chromatographic fingerprint of the diesel fuel, in monotonous repetition, for many months after the accident.

Bacteria normally present in the sea will attack and slowly degrade spilled oil. On the basis of visual observations it has been said that the oil spilled by the Torrey Canyon disappeared rapidly from the sediments. This was interpreted to mean that the action of the bacteria was "swift and complete." Our analyses, which were carried out by objective chemical, rather than by subjective observational techniques, showed the steady persistence of fuel oil that should, in principle, be even more rapidly degraded than a whole crude oil. Thus, in May 1970, eight months after the spill, oil essentially unaltered in chemical characteristics could still be recovered from the sediments of the most heavily polluted areas. By the end of the first year after the accident, bacterial degradation of the oil was noted at all locations, as evidenced by changes in the fingerprint pattern of the oil. Yet only partial detoxification of the sediments had occurred, since the bacteria attacked the least toxic hydrocarbons first. The more toxic aromatic hydrocarbons remained in the sediments.

SPREAD OF POLLUTION

For our chemical and biological work we established an unpolluted control station, outside of the area that was polluted, immediately after the accident. For a short period after the accident the sediments at this station were still clean and the organisms alive in their normal abundance and distribution. However, within three weeks, oil was found at this station and a significant number of organisms had been killed. Another control station was established twice as far from shore. Within three months fuel oil from the spill was evident at this station, and again there was a concomitant kill of bottom-living animals. This situation was repeated several times in sequence, and by spring 1970 the pollution had spread considerably from the area affected initially. At this time, the polluted area off-shore was ten times larger than immediately after the accident and covered 5000 acres (20 square kilometers) offshore and 500 acres (2 square kilometers) in the tidal river and marshes.

Another significant observation was made in the spring of 1970: Between December 1969 and April 1970, the oil content of the most heavily contaminated marine station two and one-half miles north of the original spill increased tenfold. Similar but smaller increases were observed at about the same time at other stations more distant from shore. The oil still showed the typical chromatographic fingerprint of the diesel fuel involved in the September 1969 oil spill. This and the lack of any further accident in this area suggested that oil was spreading from the most heavily contaminated inshore regions to the offshore sediments. We believe that the increase in the pollution level and the spread of oil to outlying areas are related to a transportation mechanism that we do not yet fully understand. However, the drastic kill of the animals that occurred with the arrival of oil pollution at the offshore stations showed that mortality continued for many months after the initial spill, even though no visible evidence of oil remained on the shores.

We believe these observations demonstrate that chronic oil pollution can result from a single spill, that the decimation of marine life can extend to new regions long after the initial spill, and that, once poisoned, the sea bottom may remain toxic to animals for long time periods.

DESTRUCTION OF SHELLFISH RESOURCES

Our analyses showed that oysters, soft-shell clams, quahaugs (another variety of clam), and scallops took up the fuel oil. Because of the pollution, the contaminated regions had to be closed to the harvesting of shellfish. Continuing analyses revealed that the contamination of the 1970 shellfish crop was as severe as that of the 1969 crop. Blue mussels that were juveniles in the polluted area at the time of the spill generally were sexually sterile the next season—they developed almost no eggs or sperm. Furthermore, in 1970 distant areas contained shellfish contaminated by fuel oil. Therefore, harvesting prohibitions had to be maintained in 1970 and had to be extended to polluted shellfish grounds that had not been closed to the public immediately after the accident.

It has long been common to transfer shellfish polluted by human sewage into clean water to make the animals marketable again. It has been thought that a similar flushing process would remove the oil from animals exposed to oil. Indeed, taste tests showed that the objectionable oily taste disappeared from animals maintained for some period in clean water. However, we removed oysters from the contaminated areas and kept them in clean running sea water up to six months. Fuel oil was still found in the animals by chemical analysis at essentially the same concentration and in the same composition as at the beginning of the flush period.

Thus, we discovered that hydrocarbons taken up into the fat and flesh of fish and shellfish are not removed by natural flushing or by internal metabolic processes. The substances remain in the animals for long periods of time, possibly for their entire lives. The presence or absence of an oily taste or flavor in fish products is not a measure of contamination. The reason is that only a relatively small fraction of the total petroleum product has a pronounced taste or odor. Subjective observations cannot detect the presence of the toxic but tasteless and odorless pollutants. Only objective chemical analysis measures the presence of these chemical poisons. It is important to note in this regard that state and federal laboratories in the public health sector are not generally equipped to carry out these important chemical measurements. Such tests are vital, however, for the protection of the consumer.

Thus, our investigation demonstrated that the spill produced immediate mortality, chronic pollution, persistence of oil in the sediments and in the organisms, spread of pollution with the moving sediments, destruction of fishery resources, and continued harm to fisheries for a long period after the accident. Our continuing study will assess the persistence and toxicity of the oil and the eventual ecological recovery of the area. At the present time, one and one-half years after the spill, only the pollution-resistant organisms have been able to reestablish themselves in the more heavily contaminated regions. The original animal populations there have not become reestablished. Many animals that are able to move early in their life cycles, such as free-swimming larvae, reach the polluted area and are killed when they settle on the sea bottom or in the marshes at West Falmouth.

In addition, revitalization of bottom areas probably will be hampered by oxygen depletion caused by oxygen-requiring bacteria that degrade oil (Murphy, 1970).

THE SIGNIFICANCE OF WEST FALMOUTH

Some scientists are convinced that the effects at West Falmouth are a special case and have little applicability to spills of whole, unrefined crude oils. They contend that No. 2 fuel oil is more toxic than petroleum and that therefore it has effects that would not be comparable to those of whole petroleum. We cannot agree with this view.

Fuel oil is a typical oil-refining product. It is frequently shipped by sea, especially along coastal routes, and it is spilled in accidents like those which occurred at West Falmouth and off Baja California following the grounding of the Tampico Maru in 1957 (Jones et al.).

More importantly, fuel oil is a part of petroleum, and as such it is contained within the whole petroleum. Surely, hydrocarbons that are toxic when they are in fuel oil must also be toxic when they are contained in petroleum. Therefore, the effects observed in West Falmouth are typical both for that fuel oil and the whole crude oil. In terms of chemical composition, crude oils span a range of molecular weights and structures. Many light crude oils have a composition not too dissimilar from that of fuel oil, and their toxicity and effects on the environment are very similar. Other heavier crude oils, while still containing the fuel oil components, contain higher proportions of the long-lasting poisons that are much more persistent and that include, for instance, some compounds that are potent carcinogens in experimental animals. Such heavy crude oils can be expected to be more persistent than a fuel oil, and they will have longer lasting long-term effects. Even weathered crude oils may still contain these long-term poisons, and in many cases some of the moderately low-boiling, immediately toxic compounds. In our view, these findings differ from those of other investigators principally for two reasons: Our study is based on objective measurement and is not primarily concerned with the mobile, adult marine species—the fish whose migratory history is largely unknown—or the highly resistant intertidal forms of life. We are studying quantitatively the effects of the spill on the bottom animals that cannot escape the spill or the polluted sediment and that are thus exposed to chronic pollution. Since all classes of bottom animals are severely affected by the oil, we believe that the effects on free-swimming animals should be just as drastic. The difficulty of measuring the total impact of oil on the marine life has led many to doubt the ecological seriousness of oil pollution. Our findings, extending far beyond the period when the visual evidence of the oil had disappeared, are based on objective chemical analyses and quantitative biological measurements, rather than on subjective visual observations. They indict oil as a pollutant with severe biological effects.

It is unfortunate that oil pollution research has been dominated so strongly by subjective, visual observations. Clearly, oil is a chemical that has severe biological effects, and therefore oil pollution research, to be fully meaningful, must combine chemical with biological studies. Those few investigators who are using objective chemical techniques find patterns in the environmental damage by oil that are similar to those demonstrated by the West Falmouth spill. Thus, Kolpack (1970) reported that oil from the blowout at Santa Barbara was carried to the sea bottom by clay minerals and that within four months after the accident the entire bottom of the Santa Barbara basin was covered with oil from the spill. Clearly, this is one of the most significant observations in the aftermath of that accident. A concurrent and complimentary biological study would have appreciably enhanced our understanding of the ecological damage caused by the Santa Barbara oil spill.

Sidhu and co-workers, applying analytic methods similar to those used by us, showed that the mullet, an edible finfish, takes up petroleum hydrocarbons from waters containing low levels of oil pollution from refinery outflows. In their chemical structures the hydrocarbons isolated by the investigators are similar to those found in the polluted shellfish of West Falmouth. The compounds differ markedly from those hydrocarbons present as natural components in all living organisms, yet closely approximate the hydrocarbons in fossil fuels (Sidhu et al., 1970).

Numerous results of crude-oil toxicity tests, alone or in the presence of dispersants, have been published in the literature. However, in almost all cases such tests were performed on relatively hardy and resistant species that can be kept in the laboratory and on adult animals for short time periods under unnatural conditions or in the absence of food. At best, such tests may establish only the relative degree of the toxicity of various oils. We are convinced that the exposure of more sensitive animals, especially young ones, to oil pollution over many months would demonstrate a much greater susceptibility to the damaging effects of the oil. Such effects have been demonstrated in the studies of the West Falmouth oil spill. These studies represent a meaningful field test in open waters.

Thus, we believe that the general toxic potential and the persistence of the West Falmouth oil are typical of most oils and oil products both at the sea bottom and in the water column.

CONCLUSIONS

Our analysis of the aftermath of the West Falmouth oil spill suggests that oil is much more persistent and destructive to marine organisms and to man's marine food resources than scientists had thought. With the advent of objective chemical techniques, oil pollution research has entered a new stage. Earlier interpretations of the environmental effect of oil spills that were based on subjective observation, often over a short time span, have questionable validity. Crude oil and oil products are persistent poisons, resembling in their longevity DDT, PCBs and other synthetic materials. Like other long-lasting poisons that, in some properties, resemble the natural fats of the organisms, hydrocarbons from oil spills enter the marine food chain and are concentrated in the fatty parts of the organisms. They can then be passed from prey to predator where they may become a hazard to marine life and even to man himself.

Natural mechanisms for the degradation of oil at sea exist—the most important of which is bacterial decomposition. Unfortunately, this is least effective for the most poisonous compounds in oil. Also, oil degrades only slowly in marine sediments, and it may be completely stable once it is taken up by organisms. It has been thought that many of the immediately toxic low-boiling aromatic hydrocarbons are volatile and evaporate rapidly from the oil spilled at sea. This has not been the case at West Falmouth, where the low-boiling hydrocarbons found their way into the sediments and organisms. We believe that the importance of evaporation has been overestimated.

Oil-laden sediments can move with bottom currents and can contaminate unpolluted areas long after the initial accident. For this reason a *single* and relatively small spill may lead to *chronic,* destructive pollution of a large area.

We have not yet discussed the low-level effects of oil pollution. However, a growing body of evidence indicates that oil as well as other pollutants may have seriously damaging biological effects at extremely low concentrations, previously considered harmless. Some of this information was presented in Rome at the December 1970 Food and Agriculture Organization's Conference on the Effects of Marine Pollution on Living Resources and Fishing. Greatly diluted pollutants affect not only the physiology but also the behavior of many animals. Many behavioral patterns which are important for the survival of marine organisms are mediated by extremely low concentrations of chemical messengers that are excreted by marine creatures. Chemical attraction and repulsion by such compounds play a key role in food finding, escape from predators, homing, finding of habitats, and sexual attraction. Possibly, oil could interfere with such processes by blocking the taste receptors of marine animals or by mimicking natural stimuli and thus eliciting false responses. Our general ignorance of such low-level effects of pollution is no excuse for neglecting research in these areas nor for complacency if such effects are not immediately obvious in gross observations of polluted areas.

Recent reports suggest an additional environmental threat from oil pollution. Oil may concentrate other fat-soluble poisons, such as many insecticides and chemical intermediates (Hartung and Klinger, 1970).

Dissolved in an oil film, these poisons may reach a concentration many times higher than that which occurs in the water column. In this way other pollutants may become available to organisms that would not normally be exposed to the substances and at concentrations that could not be reached in the absence of oil.

The overall implications of oil pollution, coupled with the effects of other pollutants, are distressing. The discharge of oil, chemicals, domestic sewage, and municipal wastes, combined with overfishing, dredging, and the filling of wetlands may lead to a deterioration of the coastal ecology. The present influx of pollutants to the coastal regions of the oceans is as damaging as that which has had such a detrimental effect on many of our lakes and freshwater fishery resources. Continued and progressive damage to the coastal ecology may lead to a catastrophic deterioration of an important part of marine resources. Such a deterioration might not be reversed for many generations and could have a deep and lasting impact on the future of mankind.

Since present oil-spill countermeasures cannot completely elimate the biological damage, it is paramount to prevent oil spills. The recent commitment by the U.S. to take all steps to end the intentional discharge of oil from its tankers and nontanker vessels by the mid 1970s is important. As a result of this step and of the resolution of the NATO Ocean Oil Spills Conference of the Committee on Challenges to Modern Society in Brussels, December 1970, other countries hopefully also will adopt necessary measures to halt oil pollution from ships. This would eliminate the largest single source of oceanic oil pollution. At the same time steps also must be taken to reduce oil pollution from many other, less readily obvious sources, such as petrochemical operations on shore, disposal of automotive and industrial lubricants, and release of unburned hydrocarbons from the internal combustion engine.

REFERENCES

Blumer, M., 1970. Scientific aspects of the oil spill problem. Paper presented at the Oil Spills Conference, Committee on Challenges of Modern Society, NATO, Brussels, Nov. 1970.

———, et al., 1970a. Hydrocarbon pollution of edible shellfish by an oil spill. *Marine Bio.* 5, no. 3:195–202.

———, et al., 1970b. The West Falmouth oil spill. Reference No. 70-44, unpublished manuscript available from senior author, Woods Hole Oceanographic Institution, Woods Hole, Massachusetts, Sept. 1970.

Clark, R. C., and M. Blumer, 1967. Distribution of n-paraffins in marine organisms and sediments. *Limnology Oceanography* 12:79–87.

Gruse, W. A., and D. R. Stevens, 1942. *The chemical technology of petro-*

leum. 2nd ed. Mellon Institute of Industrial Research, New York: McGraw-Hill.

Hartung, R., and G. W. Klinger, 1970. Concentration of DDT by sedimented polluting oils, *Environ. Sci. Technol.* 4:407.

Jones, Laurence G., et al. A preliminary evaluation of ecological effects of an oil spill in the Santa Barbara Channel. W. M. Keck Engineering Laboratories, California Institute of Technology.

Kolpack, R. A., 1970. Oil spill at Santa Barbara, California, physical and chemical effects. Paper presented to the FAO Technical Conference on Marine Pollution, Rome, Dec. 1970.

McCaull, Julian, 1969. The black tide. *Environment* 11, no. 9:10.

Murphy, T. A., 1970. Environmental effects of oil pollution. Presented at American Society of Civil Engineers, Boston; available from author at Edison Water Quality Laboratory, Edison, New Jersey, pp. 14–15, July 13, 1970.

Ryther, J. H., 1969. Photosynthesis and fish production in the sea. *Science* 166:72–76.

Sidhu, G. S., et al., 1970. Nature and effects of a kerosene-like taint in Mullet. Paper presented to FAO Technical Conference on Marine Pollution, Rome, Dec. 1970.

Smith, J. Wardly, 1970. Dealing with oil pollution both on the sea and on the shores. Paper presented to the Ocean Oil Spills Conference on Challenges of Modern Society, NATO, Brussels, Nov. 1970.

MILTON SHAW

Nuclear Power: Its Promise and Its Problems

There is no doubt that in recent years there has been a significant change in the criteria used to choose among the alternatives available for meeting our energy needs; although economics continues to have strong influence, other factors have assumed equal or even higher importance. Altogether these criteria are (with no order of relative importance intended):

1) Economics
2) Environmental considerations
3) Safety aspects
4) Adequacy and reliability of supply
5) Logistics
6) Public acceptance
7) Availability of technology
8) Governmental regulation
9) Dependence on foreign sources—balance of payments considerations

Nuclear power's promise and its problems can be briefly analyzed by rating it, and the other practical alternative sources, with respect to these closely interrelated criteria. It is important that this examination take into account every aspect of power generation, including mining, construction, power production and waste handling and disposal. It is our view, shared by many others, that when such a comparative evaluation is made, nuclear power, through proper engineering to compensate for its potential problems and to meet environmental standards imposed on it, clearly offers the best means for meeting the large scale electric power and other energy needs of future generations in a safe, reliable and economic manner with minimum acceptable impact on the environment and on the health and safety of the public. Let me try to briefly summarize why.

1. From an economic standpoint, sufficient experience is now available to indicate that nuclear power plants are competitive with fossil-fueled power plants. While capital costs of nuclear plants have increased faster than fossil-fired plants, utilities have continued to order nuclear plants; a major reason has been the significant cost increases and shortages of oil, gas and coal. Nuclear fuel costs, in contrast, have been decreasing in real dollar cost.

2. Production of energy by any means will inevitably have some effect on the environment. Whereas fossil fuel fired plants have yet to solve difficult air pollution problems, particularly SO_2 and NO emissions, nuclear plants produce no such pollutants. The only emissions from nuclear power plants are radioactive releases which have generally been small percentages of acceptable limits. The AEC has taken action to assure that radiation exposures are kept to levels "as low as practicable."

In regard to thermal pollution, all power plants release heat to the environment (Fig. 1), but steps can be taken to assure that heat doesn't unduly affect aquatic life. For example, utilities are moving increasingly to cooling towers and ponds.

Figure 1 Comparative heat rejection potential from various generating plant systems.

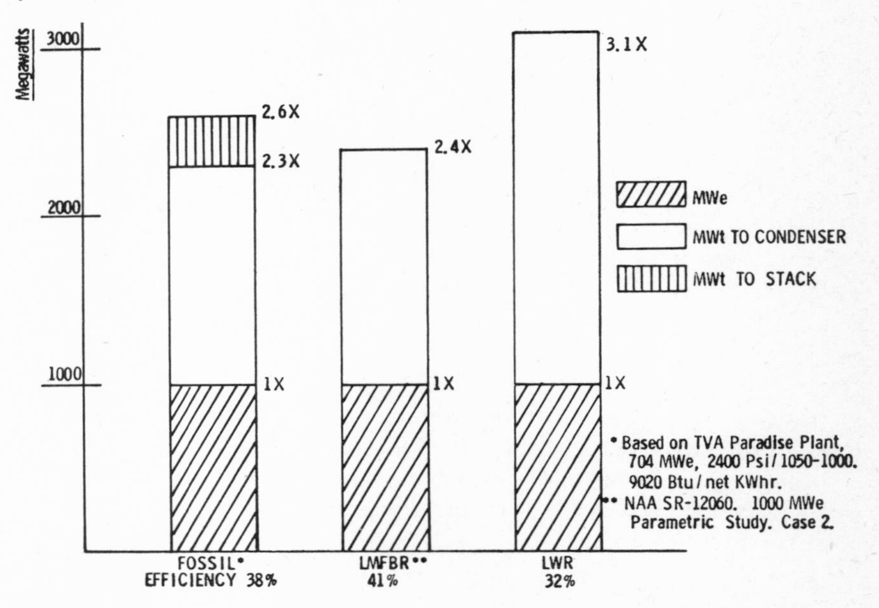

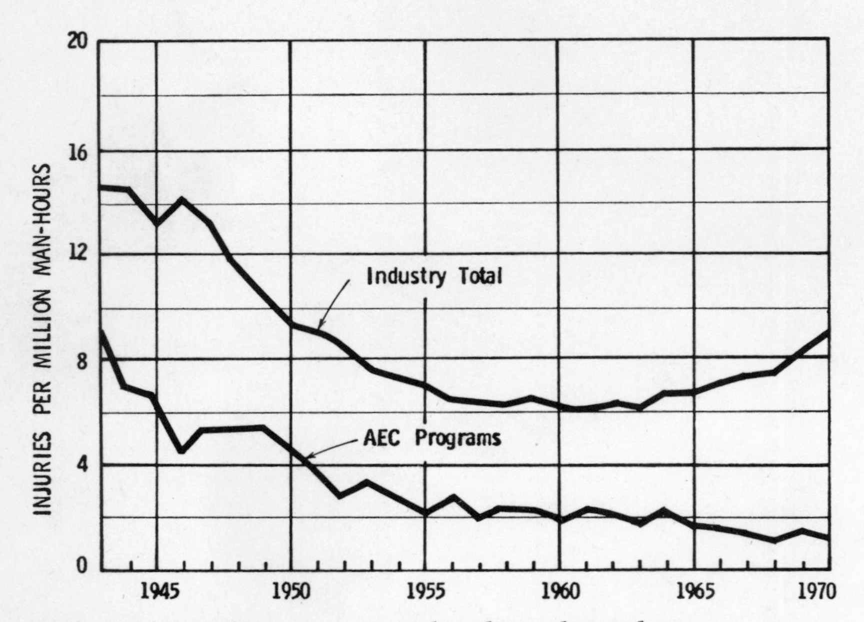

Figure 2 Injury frequency rate within the nuclear industry
Rate within the Nuclear Industry has generally averaged less than half of national rate. Only 0.4% of total man-days lost resulted directly from nuclear radiation.
Note: The injury rate for AEC programs is from WASH 1192, and does not include the uranium mining and milking industry or the operation of commercial nuclear reactors. It does include experimental R&D reactors, nuclear feed materials plants, plutonium production plants, and weapons production.

3. From its inception, the U.S. nuclear program has been conducted with extreme attention to understanding and controlling potential hazards. This program is unique in that it is the first time a comprehensive national regulation system was imposed by the government before serious accident prompted a demand for any such regulations. As a result, the safety record of nuclear power plants to date has been excellent. In the 15 years of the civilian nuclear power program, no member of the public has been killed or injured. The injury frequency rate within the nuclear industry has generally averaged less than half of national rate (Fig. 2). Only about 0.4% of total man-days lost resulted directly from nuclear radiation. This safety approach imposes a continuing requirement of strong safety and environmental research and development programs; nuclear researchers are always looking for potential problem areas.

In each of our reactor programs, we accept the fact that the remote possibility of an accident, combined with other malfunctions or deficiencies, can create a situation, which, if unabated, can lead to the release of hazardous radioactive materials. Hence nuclear power programs currently utilize three independent barriers: the fuel cladding, the primary system boundary, and the separate containment vessel.

In addition, nuclear plants are required to employ a defense-in-depth concept that has evolved in this country. This framework is expressed in terms of three levels of safety. The first level concerns the intrinsic features of the design of the nuclear plant and the quality, redundancy, inspectability, and fail-safe features of the components of the reactor and plant. The design must be such that the plant is unquestionably safe in normal operation and has a maximum tolerance for abnormal operation and component malfunction. Analyses are made to find those malfunctions or faults that could affect safety so that such malfunctions or faults can be guarded against by design, quality control, or fail-safe features as appropriate.

The second level concerns such incidents as partial loss of flow, reactivity insertions, failure of parts of the safety system, or fuel handling problems, which are assumed to occur in spite of the care taken in design, construction, and operation. Safety systems including detection instrumentation and protective devices are employed to minimize or prevent core damage in the event of such failures. Safety margins and redundancy are used in the design of the safety systems and protective features to guarantee their adequacy and reliability.

The third level concerns the postulated failure of protective safety systems simultaneously with the accident they are intended to control. Where appropriate, additional practical means may be provided to accommodate the consequences.

The safety program recognizes that the greatest importance, and therefore the highest priority, must be attached to activities related to the first and second levels because it is in these areas that the greatest safety assurance can be developed.

4. Adequacy and reliability of fuel supply currently present few problems in the case of nuclear power. Adequate uranium reserves are now available, but with projected nuclear power growth rates, we could run out of low cost uranium (less than $10/lb.) within about 30 years were we to rely only on light water reactors. However, since breeder reactors will make much more efficient use of uranium and will be relatively insensitive to its cost, they will be able to supply energy for many centuries after they are put into use.

To date nuclear power plants have achieved good reliability after initial startup, test, and maturing periods. The major problem has been to develop rapidly enough the industrial capability and facilities to design, build, and operate large numbers of plants throughout the country.

5. Once built, nuclear plants are essentially free from problems of logistics in obtaining and transporting fuel, using it, and disposing of wastes. The breeder, once it is developed, should be even more independent of fuel supply and waste disposal.

6. Concerns have been raised regarding environmental and safety aspects of nuclear power plants (and other energy systems). Steps have been taken to improve public acceptance. These include regulations for lowering radioactive releases; other improved procedures in the regulatory process (e.g., hearings on proposed rule-making); speeches and appearances by Commissioners and senior staff at symposia, public meetings and legislative hearings; and proposed siting legislation, which will permit greater public participation. Also efforts are underway to improve public understanding that practically every environmental task demands huge amounts of electrical energy, far beyond anything now available, and that while nuclear power plants are similar to other thermal power plants in terms of discharge of waste heat and electric power transmission, they can make a major improvement in other environmental considerations.

7. While further improvements are still being made, basic light water reactor technology is available. This has been well proven in the Navy program, and experience is rapidly being accumulated at 23 operable

Figure 3 Total years of operation of U.S. Central Station Nuclear Power Plants operable as of February 1, 1972.

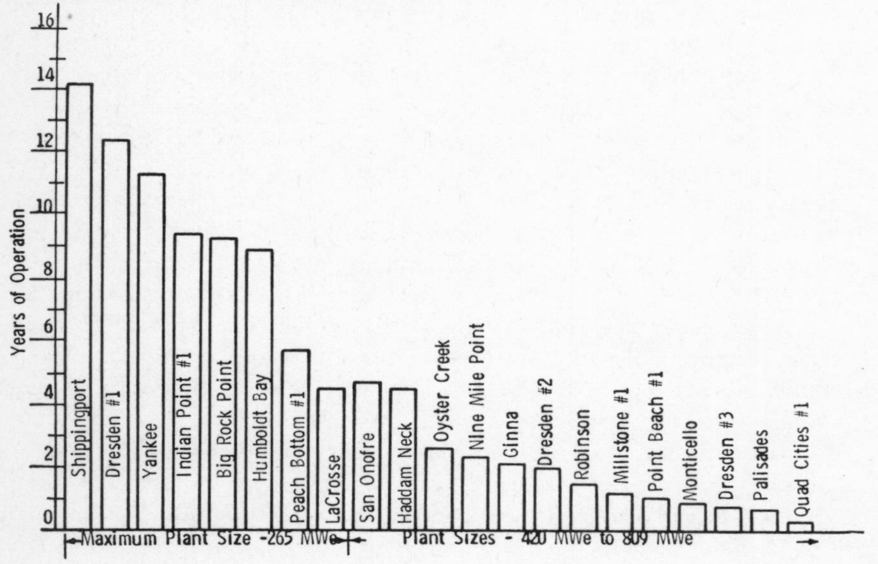

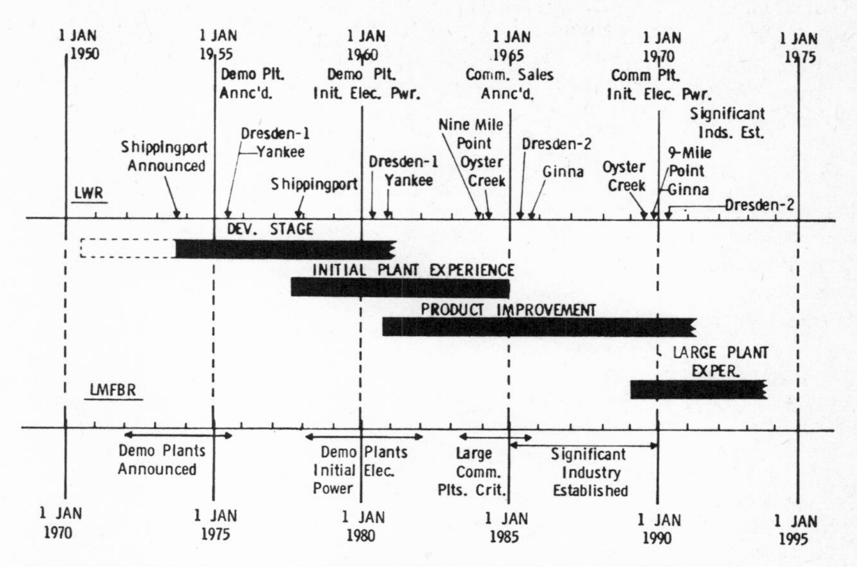

Figure 4 Development time scales

civilian nuclear power plants. It has now been about 14 years since the first civilian nuclear power plants started operation (Fig. 3). With respect to the breeder, R&D has established a sound technology base and the transition into the engineering, and the manufacturing stage has begun. The essential next step is to build demonstration plants on utility systems. Figure 4 shows the expected time scale of nuclear power development for the breeder and, for comparative purposes, that of the light water reactor.

8. Government regulation has been an integral part of the nuclear power program. Major safety and environmental reviews are conducted by the utilities, AEC regulatory staff, Advisory Committee on Reactor Safeguards, and hearings boards. Public hearings are held with liberal opportunities for public participation. The effect of the Calvert Cliffs decision was to require the AEC to look at the total environmental impact of nuclear power plants (not only radiation); the decision also required the AEC to make an overall cost-benefit judgment.

9. Finally, nuclear power has the advantage of being a domestic source of energy. Steps have been taken by the AEC to assure the viability of the U.S. uranium industry; in the future it is unlikely that the U.S. will have to depend on uranium imports as it currently relies on imports for oil. In fact, the U.S. has been exporting nuclear power technology, as shown in Table 1.

Table 1 Foreign Reactor Commitments
Operable, under Construction and Ordered

	LWR	HWR	Gas-Cooled Graphite	Light Water Cooled Graphite	LMFBR
Reactors Operable	25	10	37	8	2
Reactors under Construction	37	13	10	8	4
Reactors Ordered	28	1	—	—	—
Total Reactors	90	24	47	16	6
Total Capacity MWe	50,693	7,502	12,670	4,948	1,276

In summary, there is ample evidence to indicate that increased uses of electricity in place of other less efficient or less desirable forms of energy can not only help to avoid further deterioration of the environment, but can go a long way toward improving it. Electricity use tends to concentrate the consumption of fuel in efficient, well-controlled central power plants, and eliminates much of the widespread, relatively uncontrolled, and ineffective burning of fossil fuels in small heating or power units. Thus, while much public attention has been focused on the environmental and other siting effects of large central station power plants, it is important to keep in mind how much improvement the use of these plants represents over alternate and less effective methods of obtaining the equivalent energy.

Detailed evaluations of such alternatives have resulted in increased recognition that nuclear energy offers a new dimension to man's use of energy. Its ability to produce power with minimum environmental effects, its use in supplementing (such as by desalting) and controlling our water resources, and its help in recycling much of our wastes will be instrumental in improving our environment and conserving our most valuable natural resources. Nuclear power can be engineered to that degree deemed necessary by responsible regulatory groups at a reasonable price without giving up any of its inherent advantages. Moreover, as the nuclear program is implemented adequately, and combined with certain other important nonnuclear development efforts, major gains can be made in coping with environmental problems of energy production and reducing other types of pollution significantly. This approach provides a solid basis for a constructive and realistic course of action, and is worth pursuing in the most vigorous, disciplined and high priority manner.

J. E. MARTIN, E. D. HARWARD, D. T. OAKLEY,
J. M. SMITH, AND P. H. BEDROSIAN

Radioactivity from Fossil-Fuel and Nuclear Power Plants

Trace quantities of uranium and thorium and their products of radioactive decay are released in fly ash from large, fossil-fueled steam electric stations, raising the question of the significance of their release compared with those from nuclear power plants. This chapter presents data relative to this question that were obtained from studies conducted by the Bureau of Radiological Health at operating coal- and oil-fired steam stations and at nuclear power plants. The plants were compared by measuring or estimating exposure dose rates and calculating the fractions of the respective International Commission on Radiological Protection (ICRP) recommended limits they represented. When these fractions are related on a per-megawatt-electrical basis, it is found that the noble gases from a boiling-water reactor could produce more radiation exposure than the natural radioactivity emitted from the older, less efficient coal plant that was studied; however, for a pressurized water reactor the reverse is true. Comparisons are also made between coal-fired and nuclear power plants of current design to show the effects modern technology has on the relative radiological significance of each. The relative long-range effects of power plant fuel use are considered relative to the build-up of ^{85}Kr in the atmosphere and the release of carbon which is free of ^{14}C from fossil fuel plants. When this consideration was made, it was concluded that nuclear power reactors over the long term represent a greater overall radiological burden on the environment than fossil fuel plants, although all are well below radiation protection guides established by the U.S. Federal Radiation Council.

INTRODUCTION

Since fossil-fueled power plants discharge fly ash containing trace quantities of uranium and thorium and their products of radioactive decay, it is of interest to compare the significance of these radioactivity discharges to those from nuclear power plants. This comparison uses as a basis data that were obtained by the Bureau of Radiological Health from field studies, at operating coal- and oil-fired steam stations and at nuclear power plants (Gordon, 1968; BRH, 1969; Bedrosian, 1969; Martin et al., 1969). Prior to these studies, Eisenbud and Petrow compared the relative biological significance of fossil fuel radioactivity with emission data from nuclear plants and concluded "that an electrical generating station that derives its thermal energy from such fuels discharges relatively greater quantities of radioactive substances into the atmosphere than many power plants that derive their heat from nuclear energy" (Eisenbud and Petrow, 1964).

This comparison is not made to suggest that any radiological hazard exists for either fossil fuel or nuclear plants, but only for the purpose of showing how human exposures produced by such discharges vary by plant type and how these variations are influenced by design parameters. Likewise, there is no intent to suggest that other factors such as mining and processing of the fuel or the generation and release of other types of pollutants should not be considered. All of these factors, of course, enter into any complete determination of the total environmental burden represented; however, due to the unavailability of a common comparison index, this comparison was limited to radiological considerations only.

FIELD STUDIES

Fossil Fuel Plants

The Bureau of Radiological Health measured emissions of radioactivity and environmental levels produced by these emissions at operating fossil fuel power plants and operating nuclear power plants (Martin et al., 1969). A two-year study at oil-fired units at the Florida Power and Light Company's Turkey Point site indicated that environmental radioactivity levels produced by emitted oil fly ash were so low that detection was extremely difficult (Gordon, 1968). Since coal fly ash had been reported (Eisenbud and Petrow, 1964; Bayless and Whaite, 1965; Stone, 1967) to contain about 10 times more Ra-226 than oil fly ash and since coal plants discharge greater quantities of fly ash, studies at coal plants received more emphasis than at the oil plants. It has been estimated, for example, that a typical 1000 MWe coal plant would release about 885 times more natural radioactivity than a typical 1000 MWe oil plant (Martin et al., 1969).

The Widows Creek Station operated by the Tennessee Valley Authority was studied because it was large (1960 MWe), had older units with less efficient air-cleaning equipment, and has short stacks which would result in poor atmospheric diffusion of stack discharges. These factors suggested that environmental levels of natural radioactivity would more likely be measurable at this site than others in the TVA system and that the plant would be representive of older, less efficient plants elsewhere.

The only significant population exposure pathway at fossil fuel plants was found to be inhalation of airborne fly ash. Although long-term buildup of fossil fuel radioactivity in soil was anticipated, it was not observed. This result is attributable somewhat to the normal variability of soil and the difficulty inherent in obtaining reproductible analytical data for soil. The primary reason, however, was that specific activity of fly ash was of the same order as that of soil precluding positive determinations of fly ash in the presence of normal soil background (Martin, et al., 1969). Since fly ash is rather insoluble, the biological availability of its radioactivity is very low and any accumulation on soil was not expected to be, and was not, in fact, detectable in either vegetation or surface water above that present from the natural soil background. Positive measurements were obtained, however, from air sampling and these data and the resultant bone and lung dose rates are shown in Table 1. The air concentration data were obtained for a seven-hour sampling period during a neutral (Type D) atmospheric stability condition. The dose calculations assumed "standard man" (ICRP, 1959) characteristics.

The most significant radionuclides for exposure from fly ash were determined by examining the nuclides in the Uranium and Thorium Series relative to the maximum permissible concentrations (MPCs) for soluble and insoluble nuclides recommended by the International Commission on Radiological Protection (ICRP, 1959). If the nuclides in fly ash are soluble, bone is the critical organ; if insoluble, the lung is the critical organ. For bone the most significant nuclides in fly ash are Th-230, Th-228, Th-232, Ra-226, and Ra-228; for the lung they are Th-230, Th-232, U-238, Th-228, and Ra-228. Of these, only Th-232, U-238, and Ra-226 were actually measured; the others were assumed to be in secular equilibrium in the decay series. Thus, Ra-226 and Th-230 exist in equivalent amounts in the Uranium Series; Th-232, Th-228, and Ra-228 are present in equal amounts in the Thorium Series.

Each dose rate in Table 1 was calculated using MPCs recommended by the ICRP. For the lung exposure the MPC for unrestricted areas corresponds to a dose rate of 171 micro-rem/hr(1.5rem/yr) after suf-

Table 1 Airborne Radioactivity Concentrations and Dose Rates at the Widows Creek Plant on 13/5/69[a]

Station	Location		Air Conc. ($10^{-15}\mu$ Ci/cm^3)			Dose Rate (micro-rem/hr)[c]	
No.	Azimuth	Distance	Ra-226	Th-232	Uranium	Bone	Lung
1	34°	1.7 miles	1.3	N.D.	2.72	6.0	1.0
2	45°	1.6 miles	0.64	0.26	1.17	3.0	0.8
3	23°	3.9 miles	0.31	N.D.	2.55	1.6	0.4
4	34°	3.7 miles	0.52	0.24	2.54	2.8	0.8
5	43°	3.6 miles	0.64	0.47	1.24	3.1	1.3
6	51°	3.7 miles	0.64	0.33	4.42	3.1	1.2
7	39°	5.1 miles	0.40	0.08	1.05	1.9	0.4
8[b]	136°	2.6 miles	[b]	[b]	[b]	[b]	[b]

[a]Obtained for a seven-hour period during neutral (Type-D) stability.

[b]Station No. 8 was placed out of the plume to measure airborne dust. Air concentrations reflect subtraction of these values and filter background levels.

[c]These data are hourly exposure rates for one set of meteorological conditions and are not directly relatable to average annual exposures without accounting for varying meterorological conditions. Bone doses assumed fly ash was soluble; lung doses assumed it was insoluble.

N.D. Non-detectable above filter and normal dust background.

ficient time for an equilibrium concentration to occur in the lung tissue. For bone continuous exposure to the MPC in air for unrestricted areas for 50 years would produce a dose rate equivalent to that from a body burden of 0.01 micro-gram of radium, or 333 micro-rem/hr(0.056 rem/week). Therefore, the actual bone dose rates are very low at the beginning of exposure to airborne concentrations of fly ash radioactivity, reaching the body burden only after 50 years. For purposes of comparison, however, it is assumed that the concentrations observed at Windows Creek represent the fiftieth year of exposure, and, therefore, are a percentage of the maximum permissible dose rate recommended by the ICRP. This assuption is, of course, conservative for all exposure times less than 50 years; above 50 years the converse is true.

COMPARISON WITH NUCLEAR PLANTS

Kahn and others (BRH, 1969), conducted studies at the Dresden Nuclear Power Station to improve reactor surveillance techniques by measuring discharges and resultant environmental levels. Although taken for a different purpose, these data were readily comparable with the dose estimates made from the Widows Creek data and were used as a basis for the comparison. The most significant pathway for public exposure was from noble gases emitted from the stack. Table 2 contains dose rates for this pathway observed during neutral atmospheric stability conditions. These data are averages of several measurements made over periods of 30 to 60 minutes with a tissue-equivalent ionization chamber and Shonka electrometer. The dose rates at Dresden represented approximately the same meteorological conditions as the Widows Creek data, and since the stack heights were about the same (300 ft versus 170–270 ft), the two sets of dose rates were assumed to be comparable.

The dose limits and corresponding MPCs recommended by the ICRP for the various critical organs were used as a basis for comparing the different types of exposures. It was assumed that the ICRP values for the various modes of exposure reflected the same risk factor. The MPC for whole body exposure of individuals in the public from noble gases corresponds to 0.5 rem/yr; the MPC for long-lived bone seekers corresponds to a bone dose rate from 0.01 micro-gram of radium or 2.9 rem/yr; and the MPC for lung doses from insoluble particles corresponds to 1.5 rem/yr. Hourly exposure rates to produce these annual doses were calculated as a basis for comparing the field data.

The relative significance of the measured dose rates for the nuclear plant and those calculated from air measurements at the coal plant was determined by calculating their respective fractions of the recom-

Table 2 Measured Dose Rates in Discharge Plumes from Dresden-I

Date	Location Distance	Azimuth	Stability Condition	Release Rate	Dose Rate[a]
1/16/68	1.2 miles	190°	Neutral	14,000μ Ci/sec	32μ R/hr
1/17/68	1.1 miles	192°	Neutral	15,000μ Ci/sec	40μ R/hr
8/21/68	1.1 miles	39°	Neutral	11,600μ Ci/sec	24μ R/hr

[a]**Note:** These are time-average plume dose rates for periods of 30 min. to 1 hour; any extrapolation of these data to annual doses must account for varying meteorological conditions.

Table 3 Releases of Noble and Activation Gases from Selected Power Reactors, 1968 (10)

Facility	Type	Power	Total Discharge	Average Release Rate	Percent of Dresden-I
Dresden-I	BWR	200 MWe	240,000 Ci	7,600μ Ci/sec	100%
Humboldt Bay	BWR	69 MWe	897,000 Ci	28,200μ Ci/sec	370%
Big Rock Pt.	BWR	70 MWe	232,000 Ci	7,350μ Ci/sec	97%
Connecticut Yankee	PWR	462 MWe	3.70 Ce	0.12μ Ci/sec	0.0017%
San Onofre	PWR	430 MWe	4.75 Ci	0.15μ Ci/sec	0.0020%
Indian Pt. 1	PWR	265 MWe	55.20 Ci	1.75μ Ci/sec	0.0230%
Yankee	PWR	175 MWe	0.66 Ci	0.02μ Ci/sec	0.0003%

mended limit. These fractions were then related to each other to find which type of exposure was the greatest in terms of the recommended ICRP value. From Tables 1 and 2 the following observations can be made:

1) The maximum bone dose rate from Widows Creek was 6.0 micro-rem/hr for Station 1 or 1.8% of the ICRP value of 333 micro-rem/hr for unrestricted areas.
2) The highest dose rate for lung exposure was observed at Station 5 to be 1.28 micro-rem/hr or 0.7% of the ICRP value of 171 micro-rem/hr.
3) The average whole body dose rate above background measured from the plumes at Dresden-I was 30 micro-rem/hr which is 53% of the ICRP hourly limit for 57 micro-rem/hr (0.5 rem/yr) for whole body exposure.
4) On a per MWe basis a boiling water reactor represents an exposure dose rate that is about 120 times that from Widows Creek for soluble fly ash (bone dose) or about 300 times greater for the insoluble fly ash (lung dose).
5) These values are based on limited field data and were compared for fixed meteorological conditions. They would be substantially reduced when extrapolated to annual average exposure rates because of meterological variations.

Discharge data for selected nuclear power reactors reported for 1968 (JCAE, 1969) are shown in Table 3 to show how other operating reactors would compare to a coal plant such as Widows Creek. As a basis for this comparison, the last column in the table lists for each facility the percentage of the Dresden-I discharge rate. The percentages for the other boiling water reactors (BWR) are on the same order as Dresden-I and, on this basis, a similar relative comparison can be made. Gaseous discharges from the pressurized water reactors (PWR) are several orders of magnitude below the BWRs. The main reason for this difference is explained by the fact that PWRs have the ability to store gaseous wastes for about 60 days decay prior to discharge whereas BWRs discharge short-lived gases after only about 30 minutes holdup. The relative factors per MWe for the PWRs range from about 20 (Indian Point discharge and soluble fly ash) to about 2000 (Yankee discharge and insoluble fly ash) less than the coal plant.

DISCUSSION OF COMPARISONS

No attempts were made in these comparisons to determine the effects of the particle sizes of fly ash on the exposure dose. It was conservatively assumed that all the airborne natural radioactivity collected was in particles of respirable size. Measurements of particle sizes in filters from Widows Creek indicate that all particles were below 10 microns (Bedrosian, 1969). Solubility fractions for the radioactivity also were not determined for fly ash, a factor that has an appreciable effect on the dose produced in humans. The exposure dose rates, were, however, calculated for bone and lung (soluble and insoluble) to determine the range of this effect; the fraction of the ICRP limit for the bone dose was a factor of 2.5 higher than the lung dose fraction.

It is not entirely correct to relate Dresden measurements which were made over a period of about one hour to the measurements at Widows Creek which were for about seven hours. Each set of measurements was made, however, with great care to keep the sampling devices in the plume, and when the wind shifted away from the measurement locations, sampling was discontinued. Since both measurement times were sufficiently long to represent an average plume exposure condition, this factor was assumed to have about the same effect on each set of measurements. A building-wake effect probably occurred at Widows Creek because of short stacks which means that concentrations at points closer to the stack than the nearest point measured (1.7 miles) could have been higher. On this basis, the relative comparisons to Dresden measurements which were made at 1.1 miles would be reduced somewhat, but not enough to reverse the relative conclusion. The Dresden site, on the other hand, is located on a flat site and does not have the effect of valley channeling or aerodynamic downwash effects.

Any complete comparison of nuclear and fossil plants from a health standpoint should consider the radiation exposure significance of liquid radioactive wastes from nuclear plants and the health effects of chemical pollutants from fossil plants. These comparisons were not made because of the difficulty represented by having no relatable health-effect index. The presence of liquid radioactive wastes obviously adds some increment to the exposure dose from reactors but via a different pathway than the submersion dose rates that were considered for airborne releases.

COMPARISON OF NEW POWER PLANTS

A comparison of new coal plants with modern BWRs and PWRs was made in order to determine the relative radiological significance of the various plants utilizing current technology. These coal plants are about 1000 MWe in size and have efficient air-cleaning equipment and tall stacks (OST, 1968). The results of this comparison are shown in Table 4. The PWR data in Table 4 are based on the Connecticut

Yankee Power Plant (AEC, Dockets 50–213,) which is one of the most recent PWR reactors coming into service. The BWR data are based on Dresden-I and the listed dose rates are those obtained during field measurements adjusted to the annual average discharge rate reported for 1968 (JCAE, 1969). These measured dose rates are within a factor of two of those predicted by meteorological calculations. Dresden-I is assumed to be typical of BWRs of current design although no operating data are available for newer BWRs to confirm this assumption.

The coal plant data in Table 4 represent a typical plant that burns coal with 9% ash content which contains the same natural radioactivity as Widows Creek coal, has 97.5% air cleaning efficiency, and an 800 ft. stack with heated plume rise to about 1500 ft. The dose rates presented in Table 4 were calculated as fractions of ICRP limits per MWe, and, on this basis, the coal plant is 5 times greater that the PWR, but about 14,000 times less than the BWR. For insoluble fly ash, fractions of ICRP lung dose rates for the coal plant are about twice as large as ICRP fractions for the PWR and about 35,000 times less than those for the BWR. It should be re-emphasized, however, that BWRs in operation at this time produce exposure dose rates off-site that are only 1–3% of FRC guidelines, that the comparisons are only relative, and, that such radiation exposure rates are of negligible public health significance.

The increase in the relative comparison between new coal plants and new nuclear plants is attributed to increased air cleaning efficiency and higher stacks which provide much greater dilution of the effluent before it reaches ground level. The high stack has the greatest effect on the comparison which is based on the highest single individual exposure dose rate at ground level downwind. These stacks, however, diffuse the effluent over a larger area and thus could result in exposures to more people even though at a lower rate. In radiation protection philosophy such a practice is undesirable since the highest population dose commitment (man-rems) is related to the total risk that must be assumed. Even though these stacks may allow meeting of standards for chemical pollutants which are not necessarily subject to this concept, the resulting reduction of individual radiation exposure dose rates may be offset by an increased impact on a larger population. In order to account for the variation of the comparision due to high stacks, the hypothetical coal plant in Table 4 was assumed to discharge its radioactivity from an effective stack height of 300 ft which is the same as the BWR stack. No change was made in the PWR release conditions which assume a virtual-source, groundlevel release. The results of this comparison are shown in Table 5. When compared on this basis, the coal plant, as a fraction of ICRP recommendations, is

Table 4 Comparison of Modern Nuclear Plants with a Modern Coal-Fired Plant

	Coal Plant[a]	PWR[b]	BWR[c]
Size	1000 MWe	462 MWe	200 MWe
Stack	800 ft	200 ft	300 ft
Stack-effective	1500 ft	0 ft	300 ft
Stack Discharges			
Fly Ash	4.5×10^{9} gm/yr	—	—
Ra-Th	47.9 mCi/yr	—	—
Noble Gases	—	3.7 Ci/yr	240,000 Ci/yr
Liquid Discharges			
Fission Products	—	3.8 Ci/yr	6.0 Ci/yr
Tritium	—	1,735 Ci/yr	2.9 Ci/yr
Dispersion (Type D)	2.1×10^{-8} sec/m^3	2.5×10^{-5} sec/m^3	—
Wind Speed	8 m/sec	8 m/sec	6–8 m/sec
X-max (Type D)	35,000 m	840 m	1,800 m
Critical Organ (ICRP)	Bone	Total Body	Total Body
Dose Limit (ICRP)	333μ rem/hr	57μ rem/hr	57μ rem/hr
Dose Rate	35.2×10^{-6}	1.2×10^{-6}	8.7×10^{-2}
Fraction ICRP Dose/MWe	μ rem/hr-MWe 10.6×10^{-8}	μ rem/hr-MWe 2.1×10^{-8}	μ rem/hr-MWe 1.53×10^{-8}

[a]Coal plant data are based on 9% ash content, 97.5% fly ash removal and radioactivity levels in Widows Creek Coal.
[b]PWR data for the Connecticut Yankee plant (AEC, Dockets 50-213) are based on 1968 discharge data (JCAE, 1969). (10)
[c]BWR data for Dresden-I; dose rates were linearly adjusted from field measurements to 1968 discharge data (JCAE, 1969).

Table 5 Comparison of Modern Nuclear Plants with Typical New Coal Plant Normalized to a 300 Ft. Effective Stack Discharge

	Coal Plant[a]	PWR[b]	BWR[c]
Stack-effective	300 ft	0 ft	300 ft
Ra-Th	47.9 mCi/yr	nil	nil
Noble gases	nil	3.7 Ci/yr	240,000 Ci/yr
Wind speed	8 m/sec	8 m/sec	6–8 m/sec
Dispersion (Type D)	1.7×10^{-6} sec/m	2.5×10^{-5} sec/m^3	Unknown
X-max (Type D)	3000 m	840 m	1,800 m
Fraction ICRP Dose/MWe	8.6×10^{-6}	2.1×10^{-8}	1.53×10^{-3}

[a]Coal plant data are based on 9% ash content, 97.5% fly ash removal and radioactivity levels in Widows Creek Coal (Martin et al., 1969).

[b]PWR data for the Connecticut Yankee plant (AEC, Dockets 50-213) are based on 1968 discharge data (JCAE, 1969).

[c]BWR data for Dresden-I; dose rates were linearily adjusted from field measurements to 1968 discharge data (JCAE, 1969).

about 400 times greater than the modern PWR or about 180 times less than the BWR. As before, the factors are only relative and actual exposures are of negligible public health significance.

If long-term holdup of gaseous wastes is incorporated into future BWR designs, the results of all these comparisons will change. The design of the Shoreham Nuclear Power Station (AEC, Docket 50–322), for example, indicates that levels following the usual 30-minute holdup will be reduced an additional factor of 60 if stored for nine days. Longer holdup will produce correspondingly lower reductions. Nuclear Units 4 and 5 (AEC, Dockets 50–343 etc.) currently planned by the Consolidated Edison Company will also use BWRs which incorporate gaseous waste holdup; however, the reduction factor for these systems is not finalized in current design information.

LONG TERM EFFECTS

Both nuclear and fossil fuel plants will influence the radiation exposure doses received from radioactivity in the world's atmosphere. One way in which fossil fuel plants influence radiation doses is by the "Suess Effect." Carbon-14 exists in nature and represents a dose rate to man between 1.0 and 1.6 mrem/yr. Prior to 1900, the C-14 distribution within the carbon cycle was in a "steady state." Since 1900, however, the combustion of coal and oil has added to the atmosphere an enormous amount of CO_2 that is free of C-14 thereby reducing the specific activity of C-14 in the atmosphere and in those reservoirs in rapid exchange with it. This phenomenon, called the "Suess Effect," will reduce the dose to mankind between 1954 and 2000 by about 3 mrem assuming forecasts of electricity generated by fossil fuels are correct (UN, 1962).

The most significant long-term effect from generating electricity by nuclear power is the introduction of Kr-85 into the world's atmosphere. This phenomenon acts to increase the specific activity of the atmosphere and is a function of the amount of energy generated by nuclear means. Discharges of Kr-85 at operating nuclear power plants is low since it is retained in the fuel; however, the remaining amounts are released when the spent fuel is processed. Because of its long half-life (10.3 years) Kr-85 is building up in the world's atmosphere as the rate of nuclear generation increases. Coleman and Liberace have estimated (Coleman and Liberace, 1966) that, if current predictions for the use of nuclear power are true, the dose from Kr-85 in the world's atmosphere will be about 2 millirads/year in 2000 and about 50 millirads/year in 2060. Comparison of this situation with fossil fuel plants is difficult; however, two points are evident. First, fossil fuel plants use fuel which contains natural radioactivity, but do not create radioactivity as nuclear plants do. Second, the radioactivity discharged from fossil fuel plants is distributed locally changing the specific activity soil only a small amount; sometimes increasing it, sometimes decreasing it. The effects of nuclear plants and associated fuel reprocessing facilities, however, will be evident for very long periods of time because of Kr-85 in the atmosphere. The exposure doses will increase not only locally but world-wide unless effective removal processes are developed for Kr-85 in the gaseous discharges of fuel reprocessing plants. Such processes are under development but it is not known when they will be applicable to fuel reprocessing technology.

CONCLUSIONS

This study, on the basis of measurements and calculations at oil, coal, and nuclear power stations, has produced the following conclusions on the relative radiological significance of fossil fuel and nuclear power plants:

1) A comparison of the radiological significance of fossil and nuclear power plants is most difficult because gaseous radionuclides from nuclear plants are predominantly noble gases which produce whole body exposure, whereas radionuclides in fly ash, if soluble, are long-lived bone seekers, and, if insoluble, lead to radiation exposure of the lungs.
2) Coal plants discharge considerably more natural radioactivity than oil plants of equivalent size and design. Measurements of radioactivity levels in environmental media at the Turkey Point site show no detectable change in these media due to discharges of oil fly ash. Measurements at the Widows Creek plant, an older, less efficient coal station with short stacks and low-efficiency air cleaning equipment, showed detectable radioactivity in air only.
3) The measured or estimated dose rates were calculated as fractions of the respective ICRP recommended limits and these fractions were used for comparing the plants. When these fractions were related on a per-megawatt-electrical basis, it was found that the noble gases from a boiling water reactor can produce more radiation exposure than the natural radioactivity emitted from an older coal plant and that this coal plant can produce more radiation exposure than noble gases from a pressurized water reactor.
4) The largest variations in the comparisons were caused by the holdup time for gaseous wastes at nuclear plants and, for fossil fuel plants, the efficiency of air cleaning and the stack height.
5) For power plants of current design, nuclear power reactors represent a greater overall radiological burden on the environment than fossil fuel plants if both short-term and long-term effects are considered.

6) If the exposure dose rates determined in these comparisons are appropriately extrapolated by annual meteorological variations, it is found that fossil fuel and nuclear power plants are well below radiation protection guides recommended by the Federal Radiation Council.
7) The radiological exposure for fossil fuel plants is of negligible public health significance and, for this reason, further study of the environmental aspects of fossil fuel plants should focus on other pollutants.

DISCUSSION

H. J. DUNSTER: I would like to comment on Mr. Martin's reference to the dose rate from ^{85}Kr by the year 2000. The figure of 2 mrad/yr is for skin dose. The corresponding whole-body or gonad dose will only be about 1/30th of 1% of the natural background dose rate by the year 2000, even if no action is taken to remove krypton from reprocessing releases. I therefore regard the unqualified use of skin doses from ^{85}Kr as unnecessarily alarmist.

M. M. HENDRICKSON: I agree with Mr. Dunster that the dose shown by Coleman and Liberace is the surface skin dose. Their projected dose of 50 mrad/yr (skin dose) in the year 2060 is based on an airborne concentration of ^{85}Kr of 2×10^{-8} μCi/cm^3, an order of magnitude less than J. Z. Holland estimated. The annual whole-body dose from this concentration (2×10^{-8} μCi/cm^3) is about 0.5 mrem/yr.

J. E. MARTIN: We have also calculated the whole-body dose and skin doses from ^{85}Kr, and I agree with both Mr. Hendrickson and Mr. Dunster that the whole-body doses are low by comparison. The point I was attempting to make is that, unlike fossil fuel plants, nuclear power generation will add positive doses to the population at large. I also tried to indicate that the doses are, from a public health standpoint, insignificant. I would hardly view this as alarmist.

W. SCHIKARSKI (Chairman): Have you investigated the influence on your figures of including the releases of activity in reprocessing plants, because for each MWe produced by a nuclear power plant you must consider the whole fuel cycle?

J. E. MARTIN: No, we have not done this. We realize that many factors could be included in the comparisons. For example, the health effects of chemical pollutants from fossil-fuel plants could be the most significant consideration in terms of total health risk.

L. SAGAN: In computing radiation from exposure to plumes from a BWR and a PWR, did you assume that the dose was proportional to curies released at the stack, or did you adjust for the different mean half-lives of the characteristic radiogas mixtures from BWRs and PWRs?

J. E. MARTIN: No. We used the annual average discharge ratio reported by the operator and the AEC and normalized these rates to the same exposure conditions for an individual that were used for the fossil fuel plants and Dresden-I. On this basis we did not think that it was necessary to do half-life corrections. If, however, one were to estimate doses to a large population this would have to be done.

REFERENCES

Bayliss, R. J., and Whaite, H. M., 1965. A study of the radium alpha-activity of coal, ash and particulate emission at a Sydney power station. A paper presented at the 1965 Clean Air Conference, University of New South Wales.

Bedrosian, P. H., 1969. Radiological survey around power plants using fossil fuel. Unpublished Report, Southeastern Radiological Health Laboratory, Bureau of Radiological Health, Public Health Service, Montgomery, Ala., Nov.

Bureau of Radiological Health, 1968. Environmental effects of fossil fuel and nuclear power plants—progress report #1. Environmental Control Administration, Public Health Service; Rockville, Md., Oct.

Bureau of Radiological Health, 1969. Radiological surveillance studies at a boiling water reactor. DER-69-2, Public Health Service, Rockville, Md., Oct.

Coleman, J. R., and Liberace, R., 1966. Nuclear power production and estimated krypton-85 levels. *Radiol. Health ata and Rep.* 7, no. 11.

Eisenbud, M., and Petrow, H. G., 1964. Radioactivity in the atmospheric effluents of power plants that use fossil fuels. *Science* 144 (Apr. 17).

Gordon, J. A., 1968. Interim report of the study of public health aspects of fossil fuel and nuclear power plants. Southeastern Radiological Health Laboratory, Aug.

International Commission on Radiological Protection (ICRP), 1959. *Recommendations of ICRP on permissible dose for internal radiation.* Report of Committee II, ICRP. London: Permagon Press.

Joint Committee on Atomic Energy (JCAE), 1969. *Selected materials on environmental effects of producing electric power.* Office of Science and Technology, 1968. Committee Print, JCAE, 91st Congress, 1st Session, Superintendent of Documents, U.S. Government Printing Office, Washington, D.C., pp. 114–119, Aug.

Martin, J. E., et al., 1969. Comparison of radioactivity from fossil fuel and nuclear power plants. In *Environmental effects of producing electric power —Part I, Appendix 14,* Committee Print, JCAE, 91st Congress of the U.S., 1st Session; Washington, D.C., Nov.

Office of Science and Technology (OST), 1968. Considerations affecting steam power plant site selection. OST, Executive Office of the President, The White House, Washington, D.C., Dec.

Stone, G. F., 1967. A hazards evaluation of radioactivity in fly ash discharged from TVA steam plants. TVA, Industrial and Air Hygiene Branch, Feb., Unpublished Report.

Terrill, J. G., et al., 1967. Environmental aspects of nuclear and conventional power plants. *Ind. Med. Sur.* 36, no. 6 (June): 412.

U.S. Atomic Energy Commission (USAEC), Docket 50-213. Connecticut Yankee Nuclear Power Station. Final Safety Analysis Report. AEC, Washington, D.C.

———, Docket 50-322. Shoreham Nuclear Power Station. Preliminary Safety Analysis Report. AEC, Washington, D.C.

———, Dockets 50-342 and 50-343. Nuclear generating unit no. 5 preliminary safety analysis report. AEC, Washington, D.C.

United Nations, 1962. Report of the United Nations Scientific Committee on the effects of atomic radiation. General Assembly, 17th Session, Supl. No. (A/5216), New York, pp. 217, 249.

DANIEL F. FORD

A Reevaluation of Reactor Safety

A modern nuclear power plant after a period of operation contains a huge fission product inventory—on the order of ten billion curies. We depend on the postulated effectiveness of engineered safety systems incorporated in these nuclear power plants to mitigate fission product release into the environment in the event of abnormal reactor operation, as the release of any appreciable fraction of this activity into the environment could develop into an enormous catastrophe with lethal effects at ranges up to nearly 100 miles.

This is a report on the review by the Union of Concerned Scientists of available engineering data pertaining to a key reactor safety feature, the emergency core-cooling system (ECCS) (Forbes et al., 1971a, b). We have found that at present there is not sufficient information available to provide an objective confirmation of the ability of the presently designed emergency core-cooling system to perform its vital function. This safety system has never been tested under conditions expected to occur in the accident situation in which it would be called upon to perform. Moreover, the small number of semi-realistic experiments that have been performed to date indicate that various phenomena would exist in an accident situation that would very likely cancel the postulated effectiveness of the ECCS. This paper discusses some of these phenomena and rejects the contention of reactor manufacturers and the Atomic Energy Commission (AEC) that presently available experimental data provide a basis for confident reliance on this safety system.

LOSS-OF-COOLANT ACCIDENT

Most of the nuclear power reactors operating or planned for this decade use slightly enriched uranium dioxide as fuel and ordinary water as coolant and neutron-moderator. The uranium dioxide is in the form of ceramic pellets sheathed in thin twelve-foot long zircaloy tubes. There are thirty to forty thousand fuel rods in the core of a large modern reactor. The core is contained within a steel pressure vessel, itself situated in a large "containment" structure.

The fission products formed during reactor operation normally remain within the uranium dioxide fuel pellets where they are formed, although a percent or so of the halogen and noble gas activity escapes from the pellets into the gap between the pellets and the zircaloy tubing and into the fuel rod plenum. If a fuel rod ruptures, the free gas volume of halogen and noble gas activity within the rod is available for release outside the fuel rod. If a fuel rod melts because of deficient cooling, all of the gaseous and volatile fission products, comprising 20% of the fission product inventory, would be released from the uranium dioxide pellets. The impermeability of containing structures and the efficiency of gas treatment systems determine whether or not release of fission products into the environment will occur.

In the event of core meltdown, presently designed nuclear power plants would not be able to prevent the release of a substantial fraction of core fission product inventory into the environment. First, as portions of the molten core came in contact with residual water in the reactor vessel or containment structures, steam explosions would be expected that would be capable of rupturing the containing structures. Breach of the containment would mean that all of the gaseous and volatile fission products, approximately 20% of the core fission product inventory, would be available for transport under the prevailing meteorological conditions. The gaseous fission products would most probably be released as a cold, ground level cloud. An AEC study in 1957—which, it should be noted, is still up-to-date—set forth the methodology for estimating the lethality and range of such a cloud, which, when applied to a reactor of the size now operating and on the present refueling schedule, results in the estimate that under not uncommon weather conditions (night-time temperature stratification and a 6.5 mph wind) the release of core gaseous fission products could be lethal for 75 miles downwind of the plant in a corridor up to two miles wide, with acute and long-term human injuries extending for as far as several hundred miles. Given the proximity of presently operating and planned nuclear power stations to populated areas, core meltdown could clearly be a catastrophic event.

Since the release of appreciable fission product activity into the environment implies the melting of fuel elements, the primary nuclear reactor safety concern is whether adequate core cooling is guaranteed in all circumstances by system design. Potential cooling problems range from small leaks in primary cooling system piping to ruptures of the reactor pressure vessel—the five-story boiler that contains the twelve foot high nuclear core. The most serious accident presently considered credible in the design of light-water cooled power reactors is a double ended rupture of the largest pipe in the primary cooling system. The accident subsequent to such pipe ruptures is designated a "loss-of-coolant accident" (LOCA).

The LOCA begins with the rapid loss of system pressure and ejection of coolant from the reactor pressure vessel. Depending on the size and location of the rupture, and on the initial coolant conditions, the depressurization may cause violent flow conditions that the core cannot withstand. If the coolant is initially subcooled, as in a pressurized water reactor (PWR), blowdown phenomena may start with disruptive acoustic disturbances, and the rapid fluid accelerations and high fluid velocities may generate severe structural loads that can endanger the integrity of the primary system internals. During this blowdown or depressurization portion of the LOCA transient, the temperature rise of the zircaloy fuel cladding is determined by the redistribution of stored thermal energy in the fuel rods. Subsequently, decay heating by core fission products and the exothermic contribution of zircaloy-steam reactions become the driving forces in core heatup. The zircaloy-steam reaction—the burning of the fuel cladding in water—becomes a significant heat source as temperatures rise above 2100°F. Core heatup is quite rapid, and if no emergency cooling were provided at all, the clad temperature would increase from a normal temperature of 600°F to above 3000°F within the first minute following the pipe rupture. At this point addition of emergency cooling water would aggravate the accident because of the heat contributed by the exothermic reaction between the fuel cladding and the emergency cooling water. Thus, if a LOCA were not interrupted in time by effective injection of emergency cooling water, an uncontrolled meltdown would be in progress at the end of the first minute of the transient.

EVALUATION OF ECCS EFFECTIVENESS

To have confidence in the emergency core-cooling system, detailed information would be required on the nature and sequence of events during the LOCA transient that the ECCS would have to deal with and the capabilities of the ECCS, as demonstrated in suitable tests.

There are, however, major gaps in our general basic knowledge concerning LOCA phenomena and ECCS effectiveness. In 1967 and 1968 the AEC published several assessments of ECCS effectiveness made by Battelle Memorial Institute (Carbiener et al., 1967) Oak Ridge National Laboratory (Lawson, 1968), and an ad hoc Task Force called together to review this system (Ergen, 1967). The Battelle report stated flatly that "the effectiveness of the emergency core-cooling system is largely unknown" and proposed major research projects to obtain basic and much needed data. The ad hoc Task Force also recommended much research and development work. And the Oak Ridge state-of-the-art study concluded:

1) that the underlying assumptions of calculations made concerning core behavior in a LOCA had not been substantiated;
2) tests of the ECCS have never been performed in the environmental conditions of the dynamic pressure, temperature, and humidity that might prevail in a LOCA; and
3) so little information exists on key LOCA phenomena "that a positive conclusion of the adequacy of these systems would be speculative."

In response to such strong doubts about the postulated effectiveness of the ECCS, the AEC formulated a water-reactor safety research plan that identified "all the factors affecting the performance and reliability of ECCS" as "the most urgent problems in the safety program today." This research program plan was published in February 1970.

The research performed since the Oak Ridge, Battelle, and Task Force reports has made some important contributions to our understanding but has not yet provided engineering data that convincingly confirms the postulated effectiveness of ECCS. Indeed, perhaps the major contribution of the research to date has been to demonstrate that what were previously *suspected* problems concerning ECCS effectiveness are actually very *real* problems.

The latest ECCS state-of-the-art study was presented in July of this year by Aerojet Nuclear Corporation (ANC) which operates the National Reactor Testing Station in Idaho for the AEC and does major research on ECCS (ANC, 1971). This very recent survey still indicates the overall lack of knowledge about LOCA phenomena and ECCS effectiveness pointed to in the earlier assessments. ANC presented the state-of-the-art survey results in a convenient diagram (Figure 1), which shows an "event matrix" that sets forth basic LOCA developments as they relate to different parts of the reactor primary system. The diagram contains a coloring scheme that indicates in ANC's words, "the general level of understanding of each of the events; for example, whether or not we have a verified descriptive or predictive analysis capability." Those areas that are blackened are considered resolved. Events lightly polka-dotted are "under control" to the extent indicated. Events about which there has been "limited study" are colored grey.

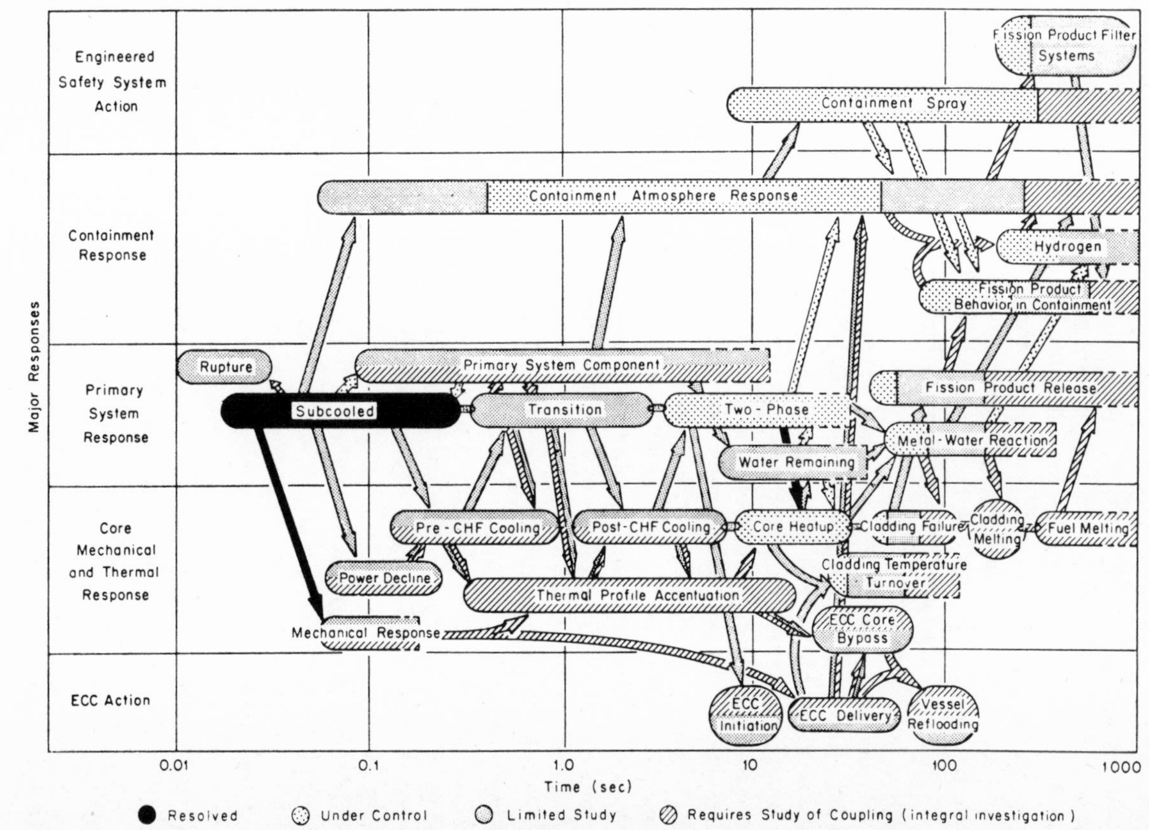

Figure 1 Major events of LOCA with engineered safety system for large break.

And events for which integral testing is required are in light hatching. The overall impression is given that only a very small percentage of LOCA transient events are well understood and that relatively unstudied and uncomprehended events abound.

Can the containment structure withstand the containment pressure transient caused by blowdown and thereby serve as a leak-tight vessel to contain fission products? This question has only received limited study. Can the core withstand blowdown forces and maintain itself in a geometry amenable to cooling? Limited study, at most, has been given to issues associated with this question. In particular, can the ECCS perform the vital function for which it was designed? This question is neither resolved nor under control; at most, certain aspects of this question have received limited study.

ANC has concluded moreover, on the basis of available data, that various mechanisms will operate during the LOCA that will cause the emergency coolant to flow around, rather than through, the areas of the core most in need of coolant or even, perhaps, bypass the core altogether. The coolant starvation produced by these mechanisms, in ANC's words, "may cancel the margin of safety previously thought to exist in emergency core-cooling systems" (Brockett et al., 1970). Research on this flow maldistribution problem was recommended in the safety research program proposed by Battelle in 1967. The problem was noted by the Oak Ridge report, the ad hoc Task Force report, and by the Committee on Reactor Safety Technology of the European Nuclear Energy Agency in their 1970 report (ENEA, 1970). And in 1970 ANC proposed a series of tests, in a document that is highly respected among reactor safety researchers, to investigate this serious problem (Brockett et al., 1970). This ANC research proposal was not funded by the AEC.

This overview of the state-of-the-art in emergency core-cooling has brought out the exceptionally limited understanding with which designers of presently operating and planned nuclear power reactors have approached what is potentially the most serious accident that can occur in such complex machines. Basic data upon which the postulated effectiveness of the ECCS is supported is fragmentary and derived from experiments whose relevance to a large modern reactor is quite indirect.

BLOCKAGE OF COOLANT FLOW CHANNELS

It is not possible to discuss all of the major uncertainties about ECCS effectiveness in this brief presentation. But to make this overall analysis of the state-of-the-art in emergency core cooling convincing to you on a technical level, I shall discuss in some detail one particular phenomenon which we expect will render presently designed ECCS ineffective; namely, the blockage of coolant flow channels in the core that arises because of fuel rod swelling during core heatup.

It was noted earlier that of the gaseous fission products roughly 1% escapes from the UO_2 fuel pellets into the gap between the pellets and the cladding and into the fuel rod plenum. This accumulation of fission product gases in the free volume of the fuel rods results in an internal pressurization of the rods which can range from 50 psi to about 2000 psi. A fuel rod's internal gas pressure is a function of the geometry of the rod and the burnup of its fuel—which is measured in megawatt-days per ton of fuel. As cladding temperature increases during core heatup in a LOCA, the strength of the cladding rapidly decreases until its ultimate tensile strength is below that required to withstand the hoop stresses resulting from fission gas internal pressure. Before the rod ruptures, internal pressure will cause swelling and ballooning of the cladding. The amount of swelling is a function of internal pressure and, to some extent, heating rate as shown in Figure 2.

Concern over rod swelling during a LOCA is obvious. If fuel rods swell on the same, or nearly the same, horizontal plane, the cross-sectional area defining the coolant passageway could be reduced so as

Figure 2 Typical fracture geometries at the indicated temperatures.

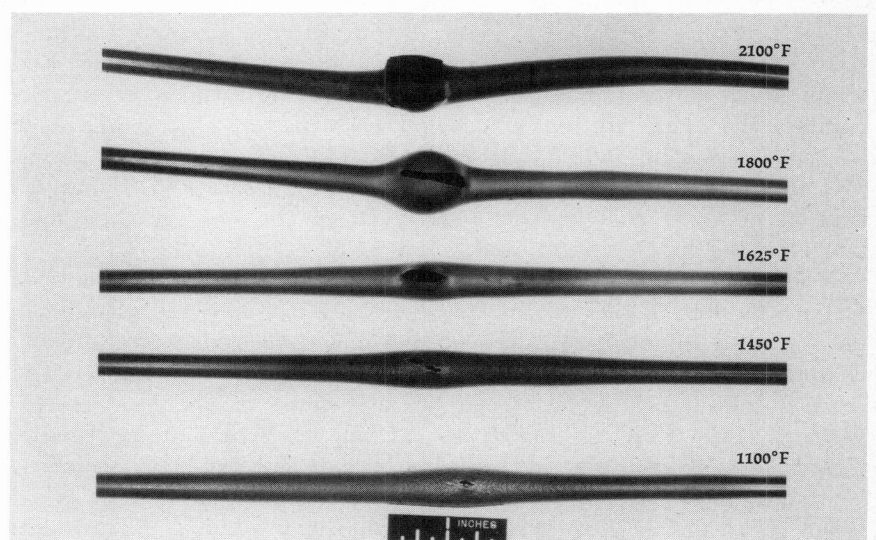

to prevent or seriously retard core cooling. Moreover, flow blockages near the center of an open-lattice pressurized-water reactor core would increase the resistance to axial flow and cause coolant to flow radially from areas of greater to areas of lesser blockage. Put another way, the flow resistance in blocked areas of the core would cause the coolant to flow around, rather than through, the region of the core most in need of coolant. Such hot-spot flow starvation would quickly initiate core meltdown.

It should be emphasized that retardation of core cooling is potentially as serious as outright prevention of core cooling. If the time at temperatures above 2000°F is long, embrittlement of the cladding can result from oxygen reactions with the cladding. In this circumstance, upon quenching by emergency cooling water the rod could fragment. If that lead to significant crumbling of the core, the debris created could develop a configuration that could not be cooled.

Because of the real and proper concern that flow blockage might prove detrimental to ECCS effectiveness, a concerted effort has been made at the Oak Ridge National Laboratory (ORNL) to study the swelling and rupture behavior of zircaloy fuel rod cladding. Experimental data evolved by ORNL has greatly clarified the determinants of rod swelling and provided data on the magnitude of flow blockage than can occur in LOCA. The reactor manufacturers have also done some work on this area, the results of which are consistent with the ORNL results.

A montage of photographs of the rupture area of tubes from one ORNL test series is shown in Figure 3. Test internal pressure increases from left to right and heating rate increases from top to bottom. Rod temperature at rupture, maximum internal pressure, and percent circumferential strain are indicated below each picture. One should keep in mind in looking at this data that AEC ECCS design criteria permit fuel tubing temperatures up to 2300°F during LOCA. With the exception of three rods in the picture, all of the rods reached their maximum swelling and ruptured below 2300°F.

If we inspect each row, we see that swelling is greater at the ends of the row than in the middle. Figure 4 shows this general relationship between rod internal pressure and the extent of rod expansion and channel blockage. This figure plots available data from single- and multi-rod tests performed by ORNL and three of the four reactor vendors. (Data from the fourth reactor vendor, Westinghouse Electric Corporation, is considered proprietary and is not publicly available.) The shape of the curve is a reflection of the changes in the stress-strain properties of zircaloy that occur during the LOCA temperature transient.

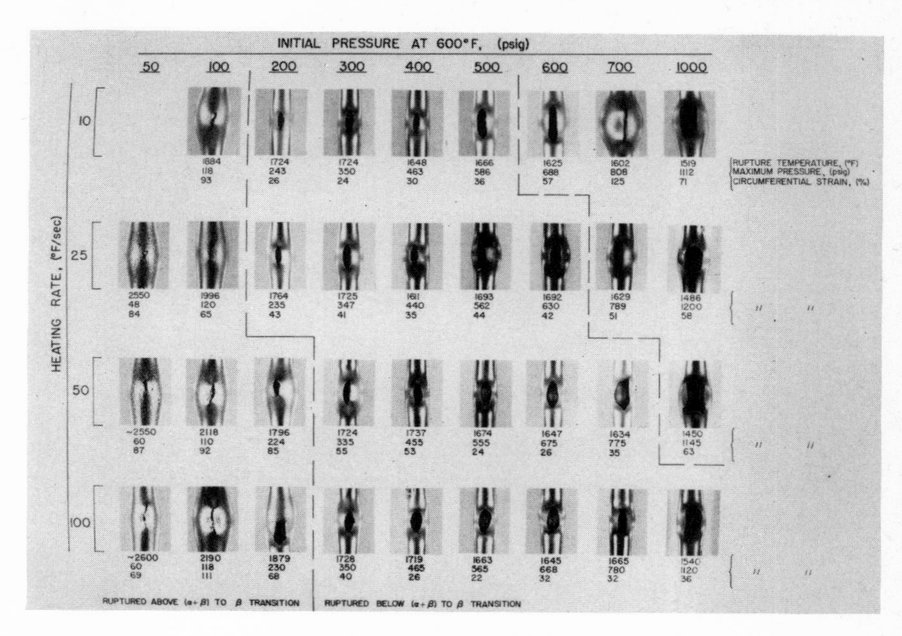

Figure 3 Rupture areas of the tubes in the base-line study.

Unfortunately, the internal pressures of fuel rods in a reactor core fall mostly in the range of pressures associated with maximum rod expansion. In a boiling-water reactor, for example, the range of internal pressures is 50 to 150 psi with most rods at less than 100 psi. Thus the experimental flow blockage data indicates the potential for development of 90 to 100% coolant channel blockages in this kind of reactor during a LOCA. One can estimate for a pressurized-water reactor as well that as much as 50% of the core could be greater than 95% blocked during a LOCA.

Studies of the disruptive effects from metal-water reaction from molten metal indicate that if any *more than a fraction of 1%* of the fuel cladding in a reactor melts that an uncontrollable situation is likely to develop. We can see that even were our estimates of the damaging level of flow blockage badly in error an accident would still present a major hazard.

The implications of flow blockages of this magnitude have not yet been explored in any systematic way. Some preliminary calculations, on the one hand, and a few experiments of doubtful relevance, on the other hand, are the only information on the effects of flow blockage presently available. The calculations, done by Battelle Memorial Institute, are presented in Figure 5. The calculation is cited by reactor

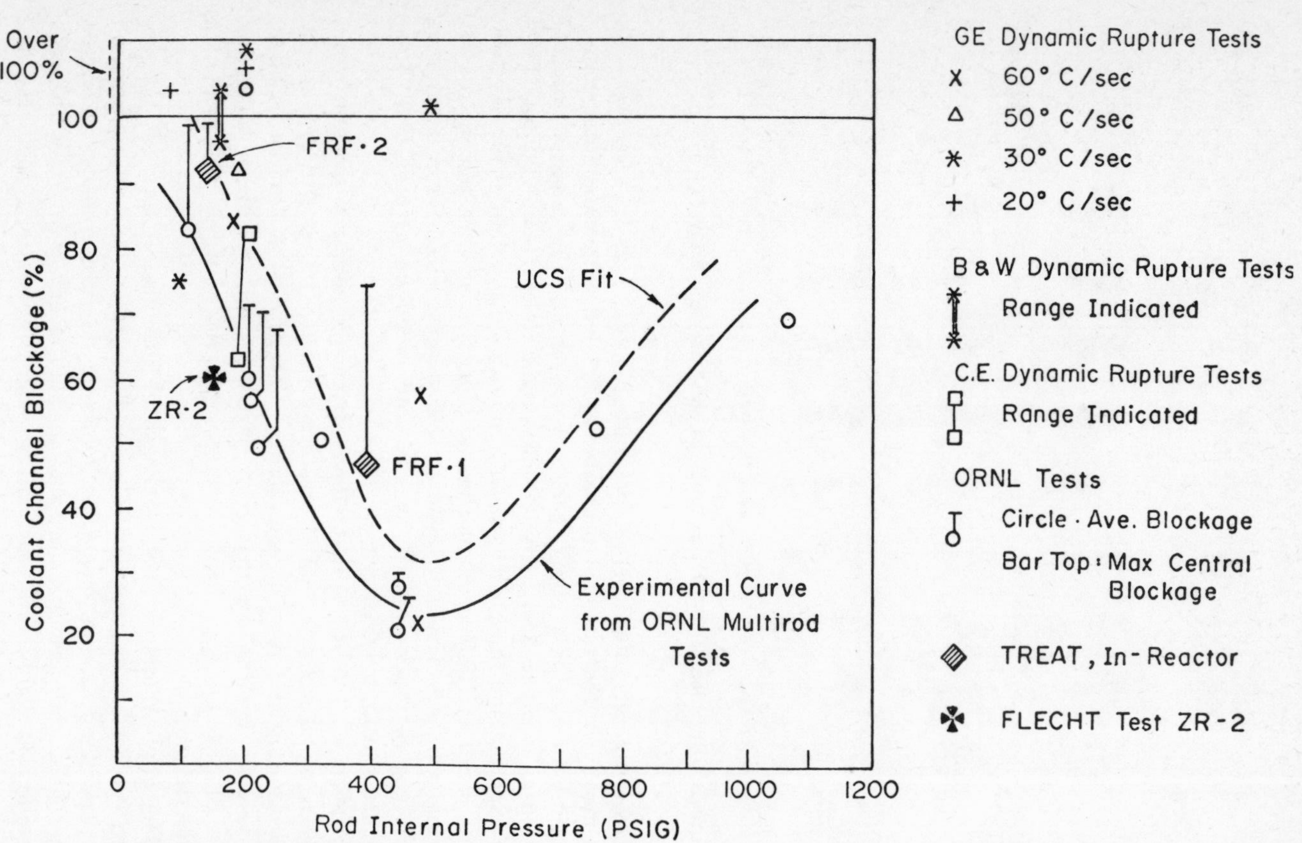

Figure 4 Single and multirod blockage tests.

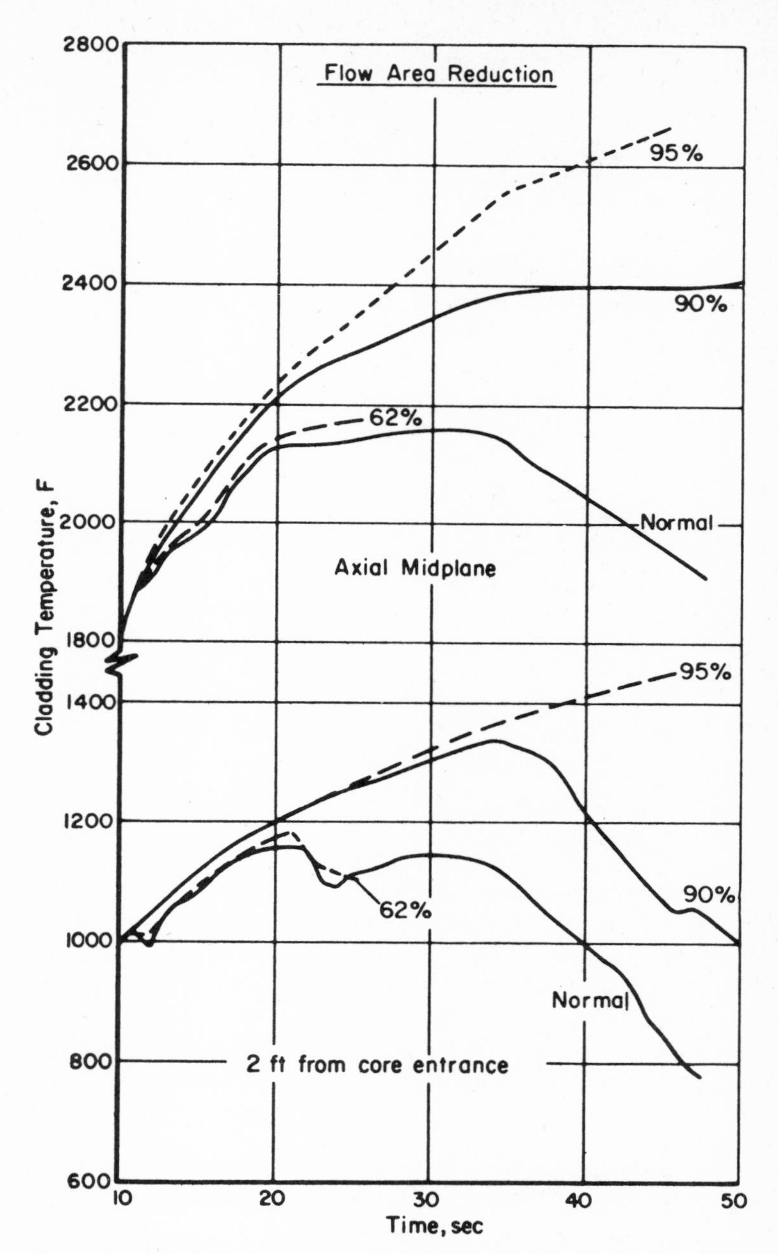

Figure 5 Sensitivity of cladding temperature history to geometric change for a maximum-power rod.

manufacturers when they seek to prove that their systems can cool cores with 50 or perhaps 90% blockage. However, the 95% or greater blockages that would be expected on the basis of available data would cause cladding temperatures to rise into autocatalytic zone for zircaloy-water reactions within 45 seconds after the beginning of the LOCA. The onset of such extensive metal-water reactions is the first step in the irreversible event of core meltdown. Even if these high blockages cause clad temperature to rise somewhat more gradually than predicted by Battelle, they would reduce cooling efficiency and tremendously increase cladding embrittlement and, therefore, the probability of core fragmentation. In short, the crude analytical models available suggest that the high blockages expected to occur in a LOCA on the basis of fuel rod swelling data could seriously degrade, if not totally negate, ECCS performance.

Experimental information pertaining to the effects of flow blockage on ECCS performance is limited to a few tests that simulated moderate flow blockage and one test that involved a full-size bundle of electrically heated internally pressurized rods. The method of simulation and degree of blockage simulated make the first set of tests essentially irrelevant to the central problem. The latter set of tests deserves more substantial comment.

The unique test of full-length electrically heated internally pressurized zircaloy clad simulated fuel rods was performed by General Electric Company under AEC contract. The test, designed Zr-2, was one of a series of 144 tests. In 143 of the 144 tests unpressurized rods were employed and thus the critical simulation of flow blockage was omitted. Test Zr-2 was deeply flawed. Twenty percent of the electric heaters in the rod bundle failed. Zr-2 did not represent a maximum power bundle. Furthermore, its instrumentation was extremely poor: for example, 57% of the thermocouples failed or performed erratically during the test; only two thermocouples were placed above the bundle midplane and both misbehaved during the test; heater rod power input was uncertain because of bungled meter and fuse arrangements. The interpretation given to Zr-2 by General Electric Company and the AEC Staff—that it demonstrated the ability of the ECCS to overcome flow blockage and effectively cool the core—extends far beyond the facts of the test and far beyond the level of reliance that could reasonably be placed on such a poorly instrumented and poorly analyzed experiment.

Since the consequences of an ECCS failure in a LOCA could be of catastrophic proportions, the Union of Concerned Scientists undertook a detailed technical assessment of this system's effectiveness, including an extensive review of available engineering data on ECCS and numerous discussions with people engaged in ECCS experimental research programs. The assurances given by the reactor vendors and by the AEC of ECCS effectiveness were not found in our review of the literature on ECCS and were not given to us by any of the experimenters on the ECCS. We found, outside of those of the reactor manufacturers, no technical reports that claimed to have assembled data sufficient to objectively demonstrate the assured effectiveness of presently designed ECCS. Moreover, other than those of the vendors, we found no expert engaged in reactor safety system research who felt that the effectiveness of the ECCS was confirmed by available engineering data. On the contrary: available engineering data not only shows ECCS effectiveness to be unproved, but indicated that there may be developments during the LOCA that the ECCS cannot overcome. Much research is planned to remove these doubts about ECCS effectiveness. Major tests of the ECCS will be conducted, on the present schedule, some time in 1975. Until this work is completed, our reliance upon presently designed ECCS is an act of faith. In view of the frailties of the ECCS emphasized in the technical literature and by those engaged in research in this area, and in view of the consequences of an uncontrolled LOCA, the Union of Concerned Scientists cannot support the licensing of additional reactors until ECCS effectiveness is assured.

REFERENCES

Aerojet Nuclear Corporation, 1971. *Presentation to atomic safety and licensing board on water reactor safety program.* Transcribed July 1.

Brockett, G. F., et al., 1970. *Simulated emergency flow effects tests (SEFET) project.* Idaho Nuclear Corporation, IN-1387, June.

Carbiener, W. A., et al., 1967. *A recommended emergency core cooling test program.* BMI-X-10208, Oct.

Ergen, W. K., et al., 1967. *Emergency core cooling.* Report of Advisory Task Force on Power Reactor Emergency Cooling, TID-24226, Oct.

European Nuclear Energy Agency, 1970. *Water cooled reactor safety.* Paris: OECD. May.

Forbes, Ian A., et al., 1971a. *Nuclear reactor safety: an evaluation of new evidence.* Cambridge, Mass.: UCS. Reprinted in *Nuclear News,* Sept. 1970.

———, 1971b. *A Critique of the new A.E.C. design criteria for reactor safety systems.* Cambridge, Mass.: UCS. Reprinted in *Nuclear News,* Jan. 1972.

———, 1972. Cooling water. *Environment,* Jan.–Feb. 1972.

Lawson, C. G., 1968. *Emergency core cooling systems for light water cooled power reactors.* Oak Ridge National Laboratory, ORNL-NSIC-24, Oct.

Union of Concerned Scientists. *An evaluation of nuclear reactor safety.* Direct testimony prepared in behalf of the Consolidated National Intervenors before the Atomic Energy Commission, Docket No. RM-50-1, Mar. 23, 350 pp.

GEORGE BERG[1]

Management of Radioactive Wastes from the Use of Nuclear Fuel

The treatment of radioactive wastes came of age as a new private industry in 1970, when the second privately owned nuclear fuel reprocessing plant under construction, and two more plants were being licensed by the AEC (Sinclair et al., 1971). The first such plant, Nuclear Fuel Services, Inc. (NFS), has been in operation in West Valley, New York, since 1966. Industry's decision to invest in the new plants was based directly on the experience gained in West Valley. In the records of West Valley we can find the story of the conception and development of a new atomic industry, a forecast of its future growth, and a measure of its impact on the environment.

THE CALL FOR A FUEL REPROCESSING INDUSTRY

Fuel reprocessing is a necessary stage for recycling nuclear fuels. The light water cooled reactors (LWR) currently used in power plants are designed to burn uranium enriched with the fissionable isotope ^{235}U. When somewhat more than three-quarters of this isotope has been used up, the fuel will not sustain a chain reaction. Before this happens, the spent fuel assemblies are taken out of the pile. In addition to uranium, each fuel element in the assembly now holds within its gas-tight cladding the waste products of the atomic chain reaction (Table 1).

Spent fuel elements are shipped to a fuel reprocessing plant, which contracts for two tasks: (1) to reclaim useable fuel and other com-

[1]This paper is based on work performed under contract with the U. S. Atomic Energy Commission at the University of Rochester Atomic Energy Project, and has been assigned Report No. UR-3490-279.

Table 1 Composition of 1 Metric Ton (1000 kg) of Fuel in a Standard Light Water Cooled Reactor (USAEC Environmental Survey, 1972, and Appendix II)

	Initial Operation		Routine Operation	
	Fresh Fuel	Spent Fuel	Fresh Fuel	Spent Fuel
Uranium-238	974 kg	959.3 kg	967 kg	947.5 kg
Uranium-235	26 kg	7.6 kg	33 kg	8.6 kg
Plutonium	0 kg	6.7 kg	0 kg	8.9 kg
Other fission products	0 kg	26.4 kg	0 kg	35.0 kg

mercially valuable isotopes, and (2) to dispose safely of the wastes. The tasks are made easier by the remarkable mass efficiency of nuclear power generation—a 1000 megawatt electric power plant spends only 1 metric ton of nuclear fuel in ten days of operation—and by the low amount of unreclaimable wastes—they are less than 4% by weight of spent fuel. Still, the tasks are difficult for a number of technical reasons: (1) nearly all the wastes are radioactive and hazardous to health, (2) some wastes are gases and hard to contain, (3) the cladding of the fuel rods may be combustible, and (4) the fuel cleaning process uses a large volume of liquids for solvents and large volumes of air for ventilation. Much of this flow becomes contaminated with radioactive materials and has to be cleaned.

Government-owned fuel reprocessing plants were already in operation in Richland, Washington, Savannah River, South Carolina, and at the Idaho Reactor Testing Station in 1963, when the AEC accepted the first bid by industry to build and run a privately-owned plant. All the factors that determined whether the investment would be commercially worthwhile were under control of the AEC. The Commission

(1) offered an already working technology
(2) assured a supply of spent fuel for reprocessing and a market for the products
(3) set the specifications for the products
(4) set the in-plant standards for the quality of operation (in terms of rules that limited the various radioactive discharges to the environment) and
(5) allowed the plant to negotiate service charges with the customers —the electric power companies.

The AEC first asked private industry to undertake nuclear fuel reprocessing in 1957, but did not succeed in making the venture attractive by business standards. Quite the contrary, a ceiling was effectively placed by the AEC on the amount of money the reprocessing plant could charge for its service, since the customer always had recourse to AEC-operated reprocessing plants. As a result, the first bid was submitted only in 1962 and it had one highly controversial feature. The contractor, Nuclear Fuel Services, Inc. (a subsidiary of W. R. Grace and Co.), offered to build and operate the plant only if allowed to cut the costs of construction and operation much below the cost of government-owned nuclear fuel reprocessing plants (JCAE, 1963).

HOW A FUEL REPROCESSING PLANT WORKS

The plant (Sinclair et al., 1971) was designed for three main functions: fuel reprocessing (double frames, Fig. 1), waste processing (single frames, Fig. 1), and long-term storage of radioactive materials (ovals, Fig. 1). Fuel assemblies were shipped by truck (a rail spur was also provided) and were uncrated on arrival and stored under water for an average of five months to let decay get rid of shorter-lived radioactive wastes. Fuel elements were then taken out, chopped into pieces, and the spent fuel was dissolved in nitric acid. The hulls of the fuel elements and all other hardware were rinsed off and sent to the burial ground. Nearly all the radioactive waste gases, such as krypton-85, were flushed into the off gas system at this point, together with water vapor carrying some of the tritium.

Uranium and plutonium salts were next separated from spent fuel by solvent extraction and ion exchange (the Purex process). Purified uranium was destined for recycling: it was suitably pretreated and packaged for storage and shipment to a uranium conversion plant. This material was easy to store because it resembled ordinary uranium obtained from ore, with low specific radioactivity.

There was no immediate civilian use for the plutonium, but it was stored for future use as a nuclear fuel. In the NFS plant, the nitric acid solution of plutonium was bottled, packaged in spillproof and radiation-shielded casks, and transferred to a directly adjacent storage facility operated by the company's landlord—the Atomic and Space Development Authority (ASDA) of New York State.

The uranium and plutonium present in spent fuel remained the property of the power company that sent the fuel out for reprocessing. The owners would eventually trade the reprocessed materials in when buying fresh fuel. Meanwhile, NFS was responsible for measuring and safeguarding every bit of fissionable uranium and plutonium from the time it was picked up at the power plant to the time it was delivered to the next user.

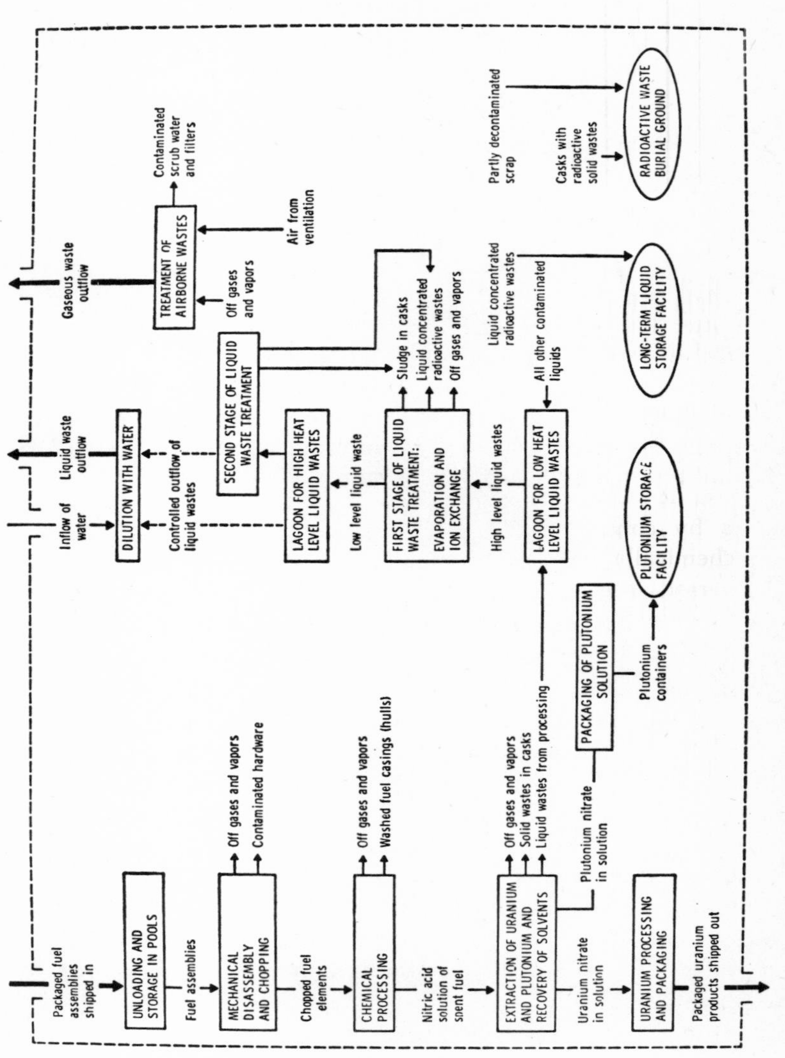

Figure 1 Operations of a nuclear fuel reprocessing plant

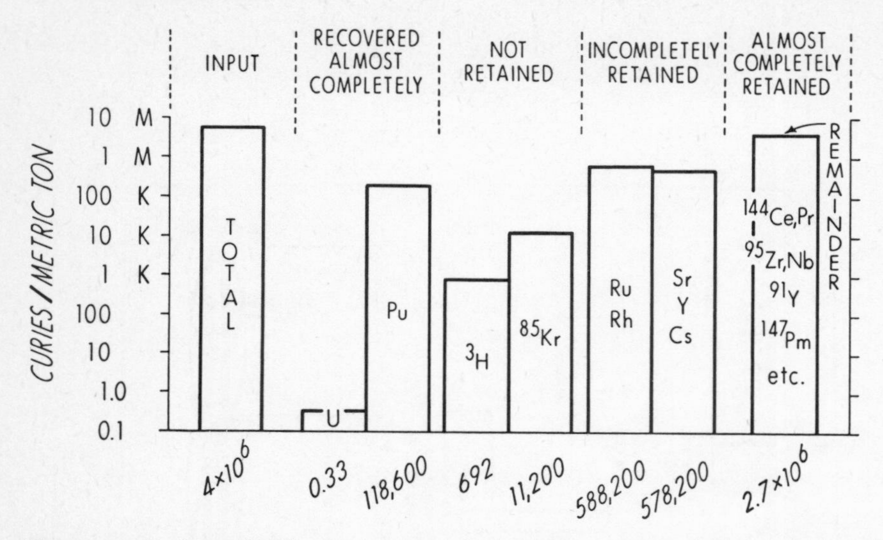

Figure 2 Budget of radioactive materials found in spent fuel during reprocessing. The data are for one metric ton (MT) of standard LWR spent fuel processed after 150 days of storage (Klement, et al., 1972). The NFS plant was designed to process up to 1 MT of such fuel per day.

The remainder of the fuel assemblies became the property—and the headache—of the reprocessing plant. The disposal of scrapped hardware and hulls was described above. Other radioactive wastes were disposed of by three routes. (1) Gases in the ventilation air stream were cleared of radioactive dusts by means of high efficiency filters, and of vapors by condensation and scrubbing; isotopes of iodine were trapped chemically; and the remaining radioactive gases and vapors were dispersed into the environment through a tall smokestack. (2) Radioactive solid wastes, such as sludges of evaporation and sediments from lagoons were rinsed, packed in casks, and buried. (3) Radioactive liquid wastes were collected in a "high-level radioactive waste" lagoon. From there on, liquid wastes were processed in stages. Each stage was designed to hold back the pollutants dissolved in water and let out cleaner water. The NFS plant was built at first with only one stage of waste cleaning (by distillation and in ion exchange). The discharge was stored in a "low level radioactive waste lagoon." The water in this lagoon was much cleaner and several thousand times less radioactive than that in the "high level lagoon," but it was still highly polluted: the "low level" name referred to heat level, and meant that the fluid was not radioactive enough to be noticeably self-heating.

After the final cleaning, the water with its remaining load of radioactive wastes was flushed through a weir into a creek, at a carefully controlled rate.

The environment of such a waste treatment plant could become contaminated with radioactive isotopes in two ways: by the intentional and controlled releases through the smokestack and the drainpipe, or by accidental releases.

Accidents came under five headings:

(1) Breakthrough at a waste outlet: for example, a filter may crack and let radioactive dust out of the smokestack, or a lagoon may overflow into a creek.
(2) Leakage from burial grounds into surface waters and ground waters.
(3) Release of airborne pollution during processing, either by a leak or by a major accident (fire or uncontrolled boiling).
(4) Release of concentrated radioactive liquids into ground water and surface runoff, either by a leak or by a major accident (ruptured tank).
(5) Pollution carried out by individuals, such as workmen with contaminated clothing or wild animals that waded through a lagoon.

The nuclear fuel reprocessing plant was designed to guard against such accidents in two ways: (1) by multiple stages of confinement and multiple stages of monitoring. For example, radioactive materials were processed in airtight and dustproof enclosures, within a series of buildings (cells) designed to contain accidental spills and blowouts within the building. The plant was fenced in (as shown by the outer frame in Fig. 1), and beyond that was a fenced-in exclusion area (Fig. 3). (2) with in-plant checks of radioactivity in the work areas and in the bodies of workers, continued with measurements of released water and air, and proceeding to routine tests of air and water beyond the plant.

THE SPARTAN DESIGN OF THE NUCLEAR FUEL SERVICES PLANT

The builder who followed this general design could still choose among plants with a wide range of performance. The state-of-the-art of handling radioactive materials was advanced enough to keep the spills in the plant and the releases into the environment far below the legal limits. By the same token, it was possible to design a plant that would perform just well enough to stay within the law.

The proposed design for the NFS plant was criticized for taking the second, cheaper option. The following is quoted from a review of the NFS proposal by a committee of experts appointed by the AEC (NFS-Lawroski et al., 1963).

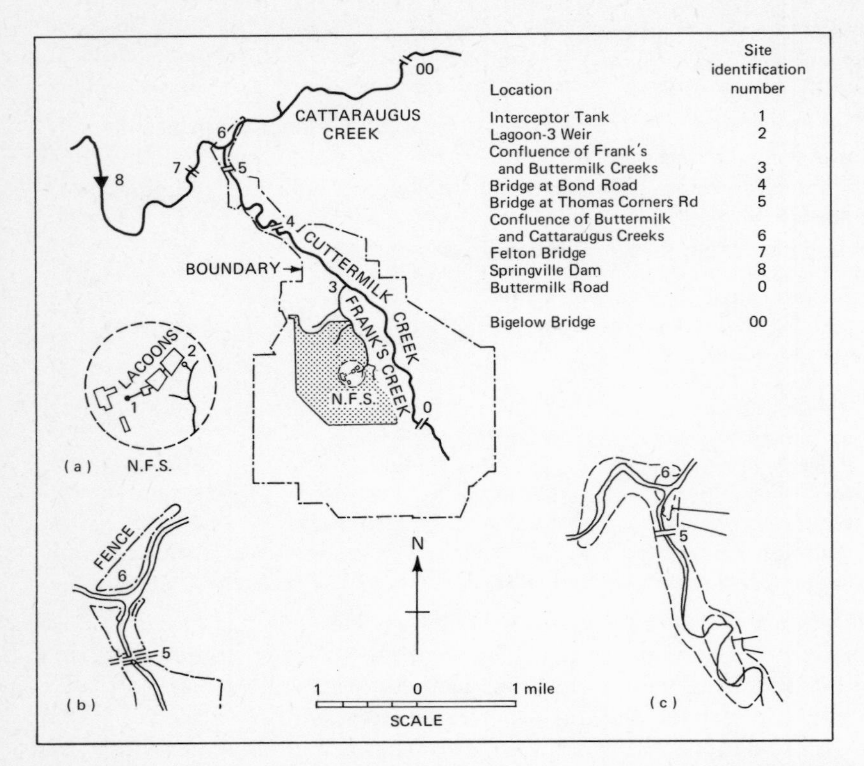

Figure 3 Map of creeks and property in West Valley, New York (Magno, et al., 1970). The insets show (a) the place where liquid radioactive wastes are released into a creek (Site 2), (b) the north end of the Service Center (Site 6), and (c) the radioactive deposits near the outlet of Buttermilk Creek, located by an aerial survey in September, 1969 (Barasch and Beers, 1972).

> . . . We are certain that new money will be required to make the facility fully meet expectations . . . in . . . marginal areas of design noted in the team's review including (1) instrumentation, (2) contamination control, (3) Purex processing, (4) waste problems, and (5) lack of laboratory space.
> . . . the proposals for manning are Spartan . . . instead of the 129 staff members proposed by NFS, 267 would be required . . .
> . . . we do not believe that this is the proper vehicle for carrying out plant process development . . .

On the opposite side of the controversy, seven electric power companies gave a grant of $2 million to NFS, and four of them contracted to become customers. New York State, through its ASDA, was a major supporter. The Agency offered a site for lease on its grounds in West Valley, and offered a number of ways to cut the costs of waste disposal.

A nuclear fuel reprocessing plant can dispose of radioactive wastes three ways: (1) store them securely on the site until the radioactivity has become negligible—which amounts to storage in perpetuity; (2) transfer them to someone else for storage and pay for the service, or (3) flush them into the environment by air or by water. The West Valley site stood on impenetrable clay, which simplified protection against leaks into ground water. The ASDA built the fuel unloading facilities and waste storage facilities for the plant, and undertook to be responsible for storage in perpetuity. The site was designed to provide ample dilution for discharges. This was of key importance, because the Federal law governing radioactive discharges into the environment (10 CFR 20) set limits to discharges, and set them in terms of concentrations of radioactive isotopes per unit volume, at the point of exit from the plant site into air or water open to the public. The more water or air the plant could use for dilution, the more wastes it could get rid of into the environment. The ASDA bought and fenced a large exclusion area—the Western New York Nuclear Service Center—to provide dilution for airborne wastes, and the entire flow of Cattaraugus Creek—over 350 cubic feet per second—was made available for flushing wastes (Kelleher, 1969). It would have been costly to divert creek water into the plant; instead, just enough land was purchased to extend the Service Center grounds to the creek (Fig. 3a). Water with radioactive wastes was actually released from the NFS lagoon into Frank's Creek, a small stream on plant grounds (the burial ground runoff also entered Frank's Creek by way of a tributary, Erdman's Brook). This emptied into Buttermilk Creek, which continued between the fences of the Service Center to its confluence with Cattaraugus Creek. A few hundred feet further downstream the fences turned away from Cattaraugus Creek. At this place, the waters of the creek were defined, in the operating license, as the discharge of liquid wastes from NFS into public waters (NFS—Logsdon and Hickey, 1971).

The plant began to reprocess nuclear fuel in October 1966. Considered as a business, NFS was apparently not a money-maker. The original owners invested some $32 million, with the expectation of a modest return of 3.6% a year, derived from reprocessing fees of $23,500 per ton of fuel (JCAE, 1963). Fees climbed to $35,000 per ton by 1972 (Larson, 1972), but the plant never posted a profit. In the 1970s, NFS was controlled by the Getty Oil Company which did not report the budget of its subsidiary to the stockholders, and there was no way to get a clear account of the gains or losses to the corporation from its venture into atomic industry (Orlans, 1967).

THE GOLDFISH BOWL SUPERVISION OF AN ATOMIC INDUSTRIAL PLANT

Technical information about the operation of NFS and its impact on the environment is wide open to public scrutiny. The environment of NFS is monitored by the Department of Environmental Conservation of New York State. Radioactive materials in environmental samples are analyzed with exemplary thoroughness at the Radiological Sciences Laboratory (N.Y. State Dept. of Health), and all the results are published quarterly. (New York State *Environmental Radiation Bulletin*). The plant and its environment are monitored by the AEC. In addition, NFS itself reports on its handling of radioactive materials to the AEC. This information is available on request (see references to West Valley documents). In practice, the reports of all the monitoring systems complemented each other: radioactivity in water was regularly sampled at the point of discharge into the creek, further downstream in Buttermilk Creek, and finally in Cattaraugus Creek at the Springville Reservoir, while airborne radioactivity was sampled at the smokestack, at the plant's boundaries, in the fallout in the general neighborhood of the plant, and finally in cow's milk in area farms (NFS—Logsdon and Hickey, 1971).

THE RECORD OF SAFETY

NFS managed to keep radioactive pollution under the limits specified in the operating license.

Accidents did not release appreciable amounts of radioactivity beyond the plant, but a few times a month something happened that contaminated the ground within the plant, the working environment, or the workers. For example, radioactive waste showed up in Erdman Brook in September 1966, when the reprocessing plant came into full operation. It was ruthenium-106, leached from the burial ground which was already in use (NFS—Kelleher, 1969; Sax et al., 1969). Later on, radioactive wastes were found leaking into Frank's Creek from a pipe that was supposed to drain only the overflow of toilet water. Some chopped and extracted casings caught on fire. A filter failed in the off-gas system, and an excessive amount of radioactive dust was discharged from the smokestack before ventilation could be shut off. Many of these mishaps were followed by changes in the equipment and procedures in the plant.

It was the normal operation of NFS, however, that exposed workers to serious amounts of radiation, and discharged radioactive isotopes to the environment. Occupational exposures to hard radiation are controlled by very detailed rules, which normally keep any worker's exposure to less than 3 rem a quarter and 5 rem a year (Wang, 1967).

One person, however, need not work in the same "hot" environment for a whole quarter, and the regulations are satisfied if he is rotated off that job just short of being exposed to the limit. This was routinely done at NFS, where a health physics staff kept tabs on each worker's accumulated exposure to hard radiation. As a result, the approximately 100 workmen in the hot areas of the NFS plant were exposed, on the average, to 7.2 rem per person in 1971. They belong to that small group of atomic industry workers—fewer than 700 in the U.S., all told—who are routinely exposed to a whole body dose of at least 4 rem of hard radiation per year (Klement et al., 1972). In addition, temporary workers were hired for one-time, unskilled jobs in contaminated areas. Their exposures amounted, as a rule, to less than 1 rem per job or 2 rems per quarter, but there were some 1400 such supplemental workers in 5½ years of fuel reprocessing at the plant. A few of the permanent employees inhaled plutonium and may have suffered a serious hazard to health. Most of the radiation doses, however, came from outside the body and did not on the balance amount to a major occupational hazard. According to the management, the principal occupational hazards to NFS workers were chemical injuries and fires associated with handling strong acids and volatile solvents.

THE PERFORMANCE OF THE PLANT AND ITS ENVIRONMENTAL IMPACT

NFS was built to process up to one metric ton (1000 kg) of spent fuel a day. When spent fuel was brought in for reprocessing, it held some fifty different radioactive isotopes (Klement et al., 1972). Some of them had very short half-lives and decayed to negligible amounts in storage. The remaining isotopes had to be separated and confined. This was done with a degree of success that differed from element to element.

The initial load of radioactivity, after the customary five months of storage, was still some 4 million Ci/MT (curies per metric ton of spent fuel) (Fig. 2). Only a minute fraction of this, amounting to some 0.33 Ci/MT, came from uranium which made up the bulk of the fuel. This was as much radioactivity as was initially present in fresh fuel, which had only uranium in it. The rest came from radioactive isotopes bred by irradiation in the atomic pile.

Uranium and plutonium were salvaged with high efficiency in the extraction step (Table 1, Figs. 2, 4). The remaining radioactive materials were put through waste treatment before release, so that less than 0.5% of the radioactivity finally escaped to the environment. This, however, was not a small amount; the releases amounted to nearly 12,000 curies for each ton of reprocessed fuel.

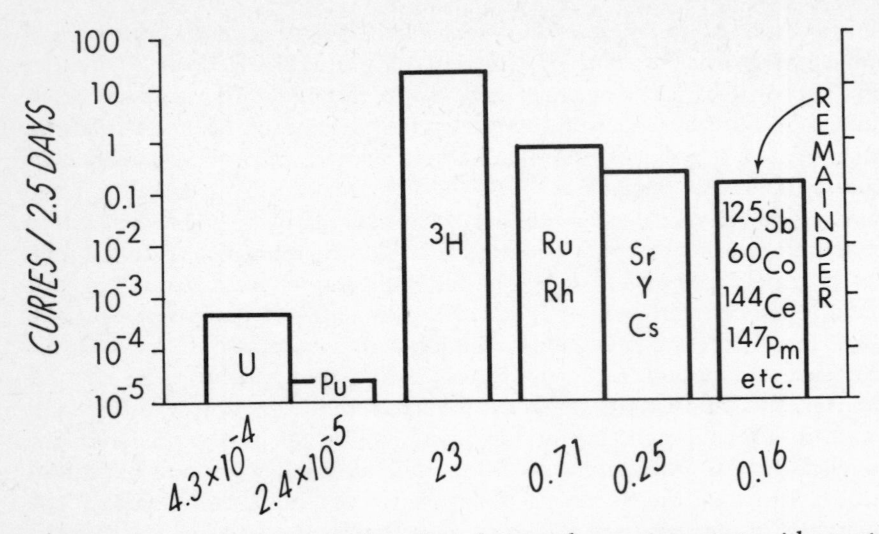

Figure 4 Radioactive materials released into the environment with waste water from nuclear fuel reprocessing. Releases from NFS during the spring and summer of 1969 are shown as the average rate of release per 2.5 days. The average amount of nuclear fuel reprocessed during that period was 1 MT per 2.5 days (Magno, et al., 1970). Figure 4 can be superimposed on Figure 2 by matching the ordinates.

The noble gases (principally Kr-85) and tritium were not retained at all. Krypton went out into the air through the smokestack and tritium was divided between releases into the creek and into the air (through the smokestack, and by evaporation of tritiated water from the lagoons). Radioactive isotopes of three other elements were incompletely retained by waste treatment and appeared in substantial amounts in the outflow of waste water. The elements were ruthenium, strontium, and cesium; the spontaneous disintegration of their isotopes also added radioactive rhodium and yttrium to waste water. The remaining elements were retained with very high efficiency (Fig. 4).

What happened to these radioactive materials after they were discharged? The outflow of air from the smokestack was held down to keep Kr-85 at less than one-half of the limit set by the operating license (the limit was 12,600 curies/day) (NFS—Davies, 1968). As a result, environmental contamination by airborne radioactive isotopes was as low in the West Valley area as it was anywhere in New York State. For example, radioactive iodine was detected only in a few milk samples; other beta-emitting isotopes (including strontium-90) were detected in milk sampled from all farms in New York State, but those amounts rose and fell with changes in atmospheric fallout from bomb tests (West Valley documents—Wilcox and Wenstrand, 1972), and were not correlated at all with the distances from the NFS smokestack to the farms (New York State—Bentley, 1971). Krypton itself was so promptly diluted in air that its concentration dropped 2000-fold on moving from 0.6 mile to 1.8 mile from the smokestack, and was barely detectable beyond that (NFS—Cochran et al., 1973).

The release of waste water over the weir was increased or decreased to match the measured flow of water in Cattaraugus Creek. The limits for radioactive pollution were set by the operating license at 300 pCi of ^{90}Sr per liter of water in Cattaraugus Creek (more precisely, 300 pCi/l was the limit for average sample readings, and 600 pCi/l for single readings). The actual releases averaged no more than 60 pCi of ^{90}Sr per liter by the time they mixed in Cattaraugus Creek, and this brought all the other waterborne radioactive isotopes well below the limiting concentrations in the same samples. There were no official limits for radioactivity in deposits on the bottom. Some radioactive materials accumulated in sand and silt of the creeks (Frank's, Buttermilk, and Cattaraugus) to be occasionally scoured out and carried further downstream by spring floods (Fig. 5). These deposits absorbed

Figure 5 Radioactivity in silt in Cattaraugus Creek, downstream from Buttermilk Creek (Wilcox and Wenstrand, 1971). The radioactive isotopes responsible for the 1969 reading of beta activity were predominantly ^{106}Ru and ^{137}Cs, with lesser amounts of ^{134}Cs and ^{90}Sr (Magno, et al., 1970).

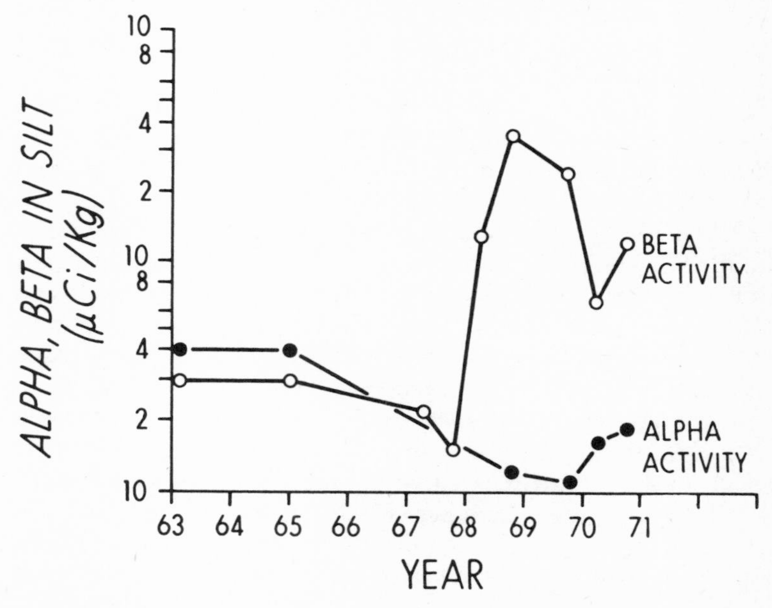

about 1/10 of the ^{90}Sr, about 1/3 of ^{106}Ru, and close to 99% of the cesium from the flow of waterborne wastes (NFS—Magno et al., 1970). When mapped from the air, the radioactivity in sand and silt could be traced for half a mile down Cattaraugus Creek; the higher concentrations shown at Site 6 in Figure 3(c) were located on a sandbar at the mouth of Buttermilk Creek.

Wild animals strayed past the NFS fences and fish swam in the polluted sections of the creeks. The first two deer taken on the NFS site in the 1967 hunting season were found to have some 70 pCi of radioactive cesium per gram of meat, together with other radioactive isotopes which they could have picked up from the water in the lagoon. After that, the fences around the NFS area were improved (New York State—Thompson, 1968; 1967). This must have kept the deer out, because deer taken near the site in all the following years were much less contaminated, seldom exceeding 2 pCi/gram of meat and usually showing a tenth of that (New York State—Bentley, 1970). Some deer, woodchucks and rabbits were found in the vicinity of the NFS plant with radioactive iodine-129 in their thyroid glands. A relatively small concentration of this long-lived isotope was released into the environment (I-129 in liquid wastes amounted to 3 units of radioactivity per 10,000 units of tritium; I-129 in the discharge from the smokestack amounted to 1 unit per 10^8 units of krypton-85) (West Valley documents—Wenstrand and Wilcox, 1972). Even this was enough to raise the concentrations of iodine-129 in the environment to levels that reached 7.7 pCi per liter in cows' milk (Daly et al., 1974). The wild animals with highest concentrations in their thyroids may have been drinking water from the most contaminated creeks (Frank's and Buttermilk), since they were also contaminated with radioactive cesium. The woodchucks, and some cattle in the area, probably accumulated traces of I-129 by eating vegetation contaminated with airborne fallout from the NFS smokestack.

Most of the fish taken from Cattaraugus Creek had the same range of concentrations of radioactive isotopes as fish from other Great Lakes regions. Some fish—especially among the bottom-feeding bullheads and suckers—did accumulate higher amounts of radioactive cesium (up to 2 pCi/mg) in the meat, and of radioactive strontium in bones (New York State—Bentley, 1970, 1971; Thompson, 1968).

What was the impact of this radioactive pollution on people? Only a few might drink the contaminated water of Cattaraugus Creek or eat an occasional meal of contaminated game and fish. The State Health Department computed the worst possible case, which was that of a diet in which all the meat came from the most contaminated deer. Even that exposure did not reach a level that was recognized as hazardous to an individual. There was, of course, a recognized hazard to a human population from any additional exposure to radioactivity. This, however, had to be computed by summing the exposures of all the exposed people. By such a standard, the hazards of local pollution by radioactive wastes from the NFS plant were much smaller and much less significant for people than the hazards of global pollution by the tritium and radioactive krypton released from the same plant.

ENVIRONMENTALISTS CHALLENGE THE PLANT

All this information was considered by the AEC's Division of Compliance, which supervised the day-to-day operation of the fuel reprocessing plant and found it to be as safe and clean as required by the license. Others, however, accused the NFS plant of polluting the environment beyond permissible limits, and not just occasionally but in daily operation. The charges published in 1968 by the Rochester Committee for Scientific Information (RCSI) and the Monroe County Conservation Council (NFS—RCSI, 1968) included an estimate that the liquid wastes discharged from the NFS plant contained 14,000 pCi of ^{90}Sr per liter on the average, and that the pollution of Buttermilk Creek with strontium-90 reached as high as 4000 pCi/l in one sample; such pollution was, according to the RCSI, far in excess of official limits for the discharge of waste water. These charges were promptly denied by the officials responsible for the compliance of NFS with the terms of its operating license. The controversy, however, quieted down in a few weeks, as it became clear to all concerned that there was no disagreement about facts. Buttermilk Creek was, indeed, some ten times more polluted than Cattaraugus (Fig. 6), just as could be expected from the greater dilution of wastes in the larger creek. The liquid wastes released over the weir were highly contaminated with radioactive materials: NFS, for example, reported that it released 1.95 million gallons of waste water in January 1968 with an average of 73,000 picocuries of ^{90}Sr per liter. The disagreement between the conservationists and the officials responsible for policing environmental pollution concerned only the rules.

The license for the operation of NFS defined the waters of Cattaraugus Creek (between Site 6 and Site 7 on Fig. 3) as the discharge from the plant. In challenging this, members of the RCSI canoed and hiked along Cattaragus Creek and dipped water from the outlet of Buttermilk Creek, unhampered by the State Preserve fences up on shore. All of Cattaraugus, they claimed, was a recreational waterway; consequently, the place where radioactive wastes were officially discharged into waters open to the public had to be somewhere closer to the actual

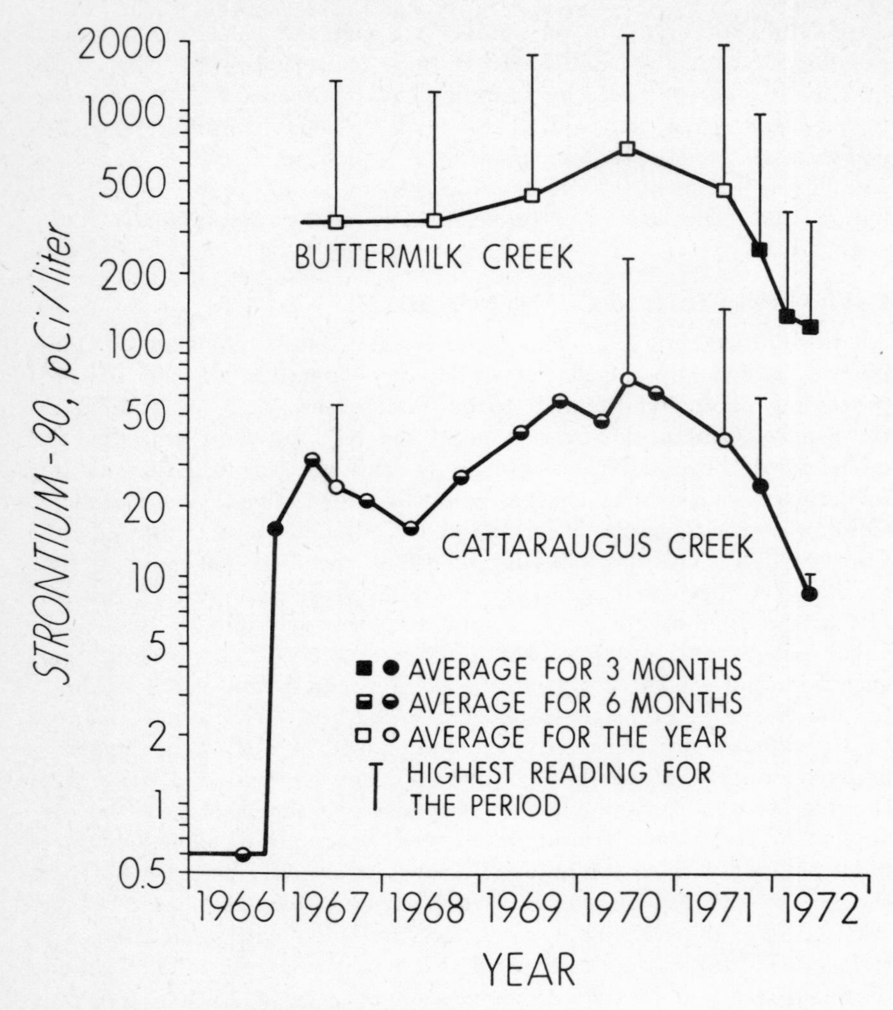

Figure 6 Strontium-90 in West Valley creeks
Points show average concentrations of ^{90}Sr in the water in units of pCi/liter (on the ordinate). Measurements were averaged on a quarterly, semi-annual or annual basis, in the years shown on the abscissa. The top of each bar shows the single highest reading among the samples used for one average. Buttermilk Creek readings are in the lower curve. Samples were taken near Site 5, Fig. 3. Cattaraugus Creek readings are in the upper curve. Samples were taken near Site 7, Fig. 3. The dashed line is at the limits set in technical specifications for average concentrations of discharged ^{90}Sr. Data from: *Envir. Rad. Bull.* 1966–1972; Thompson, 1967, 1968; Bentley, 1970, 1971; Nuclear Fuel Serv. 1972; Wenstrand and Wilcox, 1972; Wilcox and Wenstrand, 1971.

point of discharge from the lagoon into a creek. The RCSI asked for changes in the processing and monitoring of radioactive wastes at the NFS plant.

NEW GROUND RULES FOR THE INDUSTRY

At this point, the rule-making bodies within the Atomic Energy Commission stepped into the controversy on the side of the conservationists, and began to hand down a series of decisions that curbed the discharges of radioactive wastes. The following quotations tell the story.

From: U.S. A.E.C. Division of Materials Licensing
To: Nuclear Fuel Services, Inc.

"The technical specifications which are a part of the Provisional Operating License are intended to establish upper limits . . . The numerical values in these technical specifications are not intended to establish normal operating points."

"It appears impractical to restrict public access to the segment of Buttermilk Creek flowing within the NFS property boundary."

". . . significant reduction in the levels of radioactivity discharged to the watershed should be made, particularly in Buttermilk Creek, which should be considered to be a public stream." (May 31, 1968)

". . . your environmental monitoring program should be expanded. We are particularly interested in the releases to the Cattaraugus watershed, in the concentration of radioactivity in stream biota and silts, in transfer of radioactivity to human foods such as fish and deer, and in the potential resultant dose to the public." (May 27, 1970)

"You requested authorization for deep-well disposal of low-level waste from the West Valley plant. It has become clear that questions related to public health and safety raised by deep-well disposal will not be resolved in the near future . . . Thus, it is important that you develop an alternative plan . . . to comply with Appendix B, Table II of 10 CFR Part 20 for radiation in liquids, at the point of release from the lagoon rather than down stream as presently provided in your license." (May 27, 1970)

From: Nuclear Fuel Services, Inc.
To: Division of Materials Licensing, U.S. A.E.C.

"We are . . . proceeding immediately to set up the chemical treatment system . . . the first step would result in a reduction of at least 90% in the strontium and cesium discharged to the surface waters . . ." (August 14, 1970)

By the end of 1971, there were visible improvements in the discharge of waterborne wastes from the NFS plant (Fig. 6). The NFS plant received new construction funds (with the expense shared by the corporate owners and the State of New York) and added a second stage to its waste treatment system. The new waste treatment facility held back enough of the strontium and cesium to bring Buttermilk Creek below the technical specifications for radioactive pollution (West Valley documents—Nuclear Fuel Services, 1972), and the AEC revised the operating license in a way that provided for Buttermilk Creek, at least, to remain reasonably clean through 1972 (West Valley documents—Technical Specifications, 1971).

The most significant improvements, however, were made in the rules for new plants. As plans for new reprocessing plants were submitted for licensing, the AEC curtailed their discharges of radioactive liquid wastes. The reactor fuel recovery plant already built in Illinois and one of the two plants under construction in South Carolina are designed to recycle their industrial waste water (Sinclair et al., 1971). In normal operation, radioactive discharges into the environment will come out of the smokestacks, and will consist almost entirely of tritium and radioactive krypton, as in the NFS plant. New plants will also have buildings and equipment of higher quality than those used by NFS. The new "quality assurance criteria" for fuel reprocessing plants are as strict as those for nuclear power plants. Storage of radioactive wastes has also been greatly restricted. Concentrated radioactive wastes may not be stored on site for indefinite periods of time, as has been the practice. Instead, the hot liquids must be converted to solid bricks within five years, and then shipped to an AEC-operated repository (Code of Federal Regulations, 1971). Even there, storage is not "perpetual": the ultimate disposal will probably be deep underground. The growing stockpile of radioactive wastes will continue to pose some vexing problems of safety, but it will be separated from fuel reprocessing plants.

The newest of the reprocessing plants now on the drawing boards will be an expanded NFS plant at West Valley (Nuclear Fuel Services, 1973). A fabricating plant may also be built on the site, to make atomic fuel for nuclear power plants. This would cut down the hazards of storage and shipment of another highly concentrated radioactive liquid—the plutonium recovered during reprocessing. Fabricating plants dilute the plutonium by using it to enrich low-grade uranium, such as the nonfissionable uranium that is also recovered from spent atomic fuel. The product has the form of solid ceramic pellets, encased in fuel cladding and built into bulky fuel elements. The plutonium found in it is spillproof in case of a shipping accident, and not readily suited to conversion into a fission bomb in case of theft.

The old NFS plant stopped reprocessing fuel in the fall of 1972 to clear the buildings for reconstruction. Its job of "plant process development" was completed. It had pioneered an industry.

THE PERFORMANCE OF THE AEC

West Valley has been the proving ground at which the AEC and private industry developed the current standards and techniques for reprocessing radioactive wastes. Its history provides lessons about the social consequences and the environmental effects of atomic power. For the lessons to be of value, however, we have to distinguish between three levels of management: (1) the management of the fuel reprocessing industry, (2) the management of the atomic industry as a whole, and (3) the management of the production of useable energy from all sources.

1. In the narrowest view, concerned only with the reprocessing branch of the industry, the policies chosen by the AEC come off badly. A business spokesman could damn the AEC for bureaucratic meddling detrimental to private enterprise, while conservationists could and did use the same record to damn the AEC for favoring private business at the expense of the environment and the workmen.

West Valley was, on the record, a remarkable show of efficiency by a commercial operation, which was rewarded with little but hard knocks. The cost of reprocessing irradiated nuclear fuels was cut nearly in half when business took the job over from the government. Nuclear Fuel Services, Inc., contracted to do the job at an appropriately low price, and did it while living up to the conservative standards of safety and cleanliness written into the operating license. The AEC, on the other hand, kept interfering to change the license and to add to the costs after the contracts have been signed. Was this a fair use of government powers?

To the conservationists, the terms of the license issued by the AEC appeared unduly lax all along. The kind of environmentally clean operation which is required of the newest fuel reprocessing plants was technically feasible in 1963, when the West Valley plant was under design. In plant, West Valley workers were subjected to higher exposures to radiation than workers on comparable jobs in government installations. The license consequently allowed costs of production to be cut at the expense of added environmental pollution and industrial exposure to hard radiation. What were the corresponding benefits to the public? At best, they would appear as savings on the price of electricity, which could be estimated as follows.

A nuclear power plant delivers electricity to the transmission lines at a cost of about 1.2¢ per kilowatt hour. Of this, only 0.02¢ (less than 2%) pays the costs of fuel reprocessing and waste storage (Larson, 1972). The consumer's electric bill usually amounts to some 3¢ per kwh because it also covers the costs of transmission and service, so that the cost of nuclear fuel reprocessing amounts to less than 1/2% of the retail price of electricity.

Enforcing the most stringent standards that were practicable in 1967 would have cost the consumer up to half a penny more on the dollar. That would have paid for controlling the pollution of the environment by more waste processing and storage, rather than by dilution and flushing; and for controlling exposures of workers by higher quality engineering, rather than by rotation on the job. Was the AEC at fault in delaying that environmental bargain until 1972?

2. The policies of the AEC appear in a clearer light when related to the growth of the whole atomic industry. The problem facing the AEC was not simply one of transferring a technology from government to private ownership. Rather, a commercial market had to be created, and a new industry had to be developed to take over the job of reprocessing nuclear fuels. This was done by the AEC in the fifteen years, counting from 1957, when the first proposals were made to private investors, to 1972, when several reprocessing firms competed in offering services to their customers.

The market, however, was more a hope than a reality in 1962, when the license for the West Valley plant was being written. The AEC was offering nuclear technology to the electric power industry, but only two nuclear power plants were in commercial operation. The decision to use uranium fuel rather than coal or oil was not yet made by the managers who were about to build a new generation of powerplants. By keeping the cost of reprocessing down, the AEC was cutting some 2% off the costs of producing electricity. This helped to tip the plants in the U.S. to nuclear fuels rather than fossil fuels. We need to ask whether the overall environmental impact of this choice of fuel was for better or for worse.

3. When nuclear fuels are examined side-by-side with other fuels used for the production of electricity, the AEC is found in a position of leadership in managing and curbing the releases of waste products into the environment. West Valley was the critical point for curbing the release of wastes from nuclear fuel, and the control measures at West Valley were far ahead of corresponding measures for the control of wastes from fossil fuels. They included:

detailed monitoring of every kind of waste by the plant operator, with the results open to the public;

independent, multi-stage monitoring by a state agency, with prompt publication of all results;

licensing procedures, in which the builder had to provide against accidental spills into air and water, and publish the plans;
policing of pollution by a federal agency with strong powers of enforcement, and with an open record of all control measures;
strict standards of environmental cleanliness, which included accounting for all waste products of the fuel cycle.

The standards for the releases of radioactive wastes in West Valley were initially based on public health standards for permissible exposures to ionizing radiation. Later, the releases of most of the isotopes were curbed further, to the limit of best available containment. By contrast, the standards for the releases of waste products from fossil fuels were still set in 1972 at levels that were associated with measurable impairments of human health and visible degradation of the quality of air and water. Monitoring and enforcement lagged behind the admittedly slack standards. The Environmental Protection Agency was not scheduled to issue any "best available technology" orders for pollution control in fossil fuel industries before 1985.

In many respects, therefore, the control of pollution from the nuclear fuel cycle provided a model for regulations that would ensure cleaner uses of all kinds of fuels in the future. In turn, the introduction of such regulations under the National Environmental Policy Act of 1970 enhanced the competitive position of nuclear fuels and made it economically possible to put still tighter curbs on radioactive pollution, as shown in the story of West Valley.

CONTROL TRENDS AND ENVIRONMENTAL NEEDS

1. Involving the Public in the Control of Pollution

As the operation of fuel reprocessing plants in the U.S. is projected beyond 1973, the hazards from unintentional releases of radioactive wastes show up only in the immediate neighborhood of the plants and come mainly from airborne I-129 and plutonium dust. The resulting exposures can be evaluated from the experience at NFS. For example, if someone lived as close to the NFS plant as the fence allowed, he would be inhaling some plutonium from plant operations—not quite as much again, as he would inhale from breathing atmospheric fallout anywhere in the U.S. (This fallout consists of the remnants of pollution from past bomb tests.) If the new plants prove cleaner than the West Valley plant in operation to the same extent that they are in design, then no radioactive isotope will be discharged in large enough amounts to damage a population of wildlife, or to contaminate food consumed by people. The achievement of this level of cleanliness can be credited to two successful innovations: one in public supervision and the other one in technology of waste processing.

In supervising environmental pollution in West Valley, New York State agencies monitor and measure every kind of radioactive waste product. All the measurements appear promptly in a bulletin that is distributed on demand and reaches local conservation organizations and science information groups (New York State—Environmental Radiation Bulletin). These groups keep the local public directly involved in policing its own environment, and ensure a continuing vigilance against radioactive pollution.

The technological advance in pollution control by the newest fuel reprocessing plants consists of separating the radioactive process streams from the bulk flows of ventilating air and cooling water. Radioactive waste water is to be recycled. This will curtail all intentional releases of radioactive ruthenium, cesium, strontium, and related waterborne pollutants.

So far, however, the two kinds of control measures have not been put together. It remains for the rebuilt West Valley plant to recycle wastewater, and for the AEC to require that Illinois and South Carolina monitor the environment as sensitively, and bring the results before the local public as effectively, as New York has been doing.

2. Controlling Global Pollution by International Agreement

The continuing production and discharge of tritium and of krypton-85 in the nuclear fuel cycle has a global environmental impact. Radioactive hydrogen gas, tritiated water and krypton gas become rapidly mixed worldwide in the atmosphere (with some delay in reaching the southern hemisphere) and in the waters (with a long delay in mixing beyond the thermocline). The resulting exposures to people are shown in Figure 7, in the curve labeled "Other Environmental Doses." This curve was computed on the most pessimistic assumption that the use of atomic power will grow rapidly, and that nothing will be done to confine ^{3}H or ^{85}Kr. Even so, the doses are insignificantly low: they are of the order of 0.1 units, which can be compared to an increase in dose from natural radiation of 10 units for a person moving from Pennsylvania to New York State, and of 115 units for a person moving from New York to Colorado. Such whole body doses may give misleading estimates of safety for some radioactive elements such as plutonium or iodine, which form hot spots in the body; they do, how-

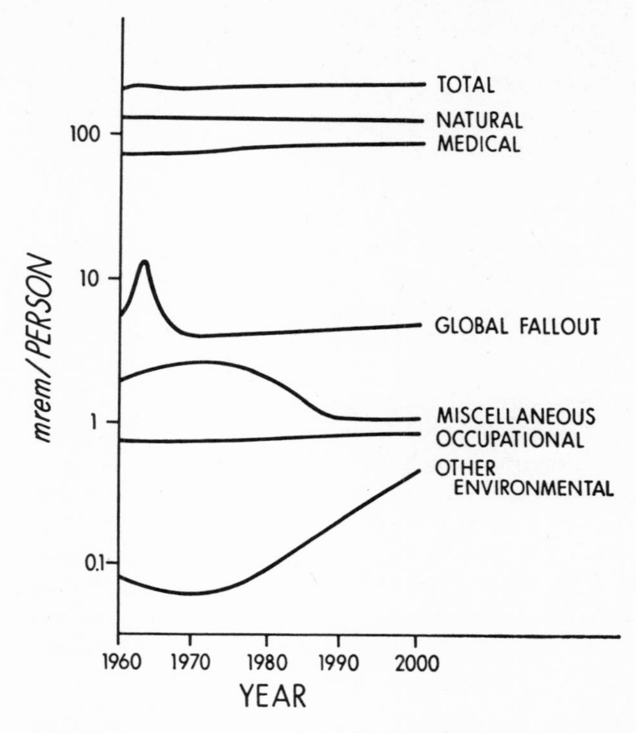

Figure 7 Exposures of people in the United States to radiation from all sources (Klement, et al., 1972). The average whole body dose of ionizing radiation, computed in mrem/person, is plotted for the years 1960 through 2000. The curve for "other environmental sources" represents mostly exposures to materials released by atomic industries.

ever, give a fair measure of the hazards of tritium and krypton-85. Consequently, they show scant reason for controlling these radioactive wastes.

In this case, however, human health does not give an appropriate measure of environmental impact. The projected releases of tritium and krypton are large enough to change the isotopic composition of elements of the whole planet: the natural tritium content of atmospheric hydrogen would be doubled before 1990 (Peterson et al., 1969), and the naturally non-radioactive krypton in the atmosphere has already been so enriched with ^{85}Kr that people working with krypton gas have to be protected from exposure to radiation. We cannot presume at this time to assess the ecological or geochemical consequences of such changes, but we can try to control them. ^{3}H and ^{58}Kr can be confined with the use of currently available technologies. Tritium is already being separated and stored at AEC's Savannah River plant and at the Commisariat à l'Énergie Atomique plant at Marcoule. Krypton can be separated from off gases by a low-cost solvent extraction process; the cost will be reimbursed at least in part by savings in plant design and operation, once the waste gases have been cleaned up (Nichols and Binford, 1971). The newest reprocessing plants, such as Barnwell, are designed to be compatible with the processes for removing krypton and tritium. The limiting factor, in this case, is the lack of international regulation of these radioactive wastes (OECD, 1972). An agreement with the force of a treaty is needed to control the releases of pollutants from the fuel reprocessing plants already operating in Western Europe (four), India (one) and Eastern Europe. Control of releases within one country—such as East Germany, which has already enacted a krypton control law—will not solve the global problem. The machinery for policing an international agreement is already in existence in the International Atomic Energy Agency, which checks reprocessing plants to safeguard fissionable materials against diversion from registered peaceful uses (NFS—Suda, 1970).

3. Marketing Cleanliness

The industry that reprocesses irradiated fuel is a secondary materials recycling industry, akin to scrap metal yards, and a waste treatment industry, akin to sewage treatment plants. The two functions have to be evaluated separately, because while West Valley is a model of good recycling, it also shows the faults that are common to all waste treatment.

The market for uranium and plutonium has been so set up by the AEC that virgin materials from the mine cannot compete with recycled materials from reprocessing. As a result, it is profitable to recover plutonium with high efficiency, and every kilo of recovered Pu eliminates the mining, processing and enrichment of 140 tons of high-grade ore. This marketing policy is in sharp and favorable contrast to the policies of the U.S. Departments of the Interior, of Transportation, and of Agriculture, which use direct and indirect subsidies to promote virgin ores over scrap metals, and mined fertilizers over recycled sewage.

There is, of course, an environmental cost as well as a benefit in substituting recycling for open-ended exploitation. In the case of nuclear fuels, recycling replaced uranium-235 with plutonium, which has a narrower margin of safety in a reactor. Plutonium-fueled reactors will have to be built at a higher level of quality control and operated at a higher level of precision to match the safety of uranium-fueled reactors. This will add to the burdens of the management of power plants, which will be forced even closer to spacecraft-type, zero-defect operation (Doub, 1972). The estimates of the hazard of accidents at power plants will have to be increased to match.

The market for radioactive waste treatment services was unconventional only in that the treatment facilities were owned by fuel makers. In other respects, the radioactive waste treatment plant had to cope with the same pitfalls that were met by any sewage treatment plant. The customers who paid the bills were the makers of wastes. They paid to be relieved of the penalties imposed on polluters by law and by adverse publicity. This put the operator of a treatment plant in the unloved position, in which all cultures put their handlers of filth. He was under pressure to deliver only the minimum of treatment necessary to forestall penalties and to keep the entire operation out of sight of the public. The pollution of a stream by wastes—whether radioactive, municipal or industrial—was pressed in such a market to a maximum, limited only by the intervention (if any) of law enforcement agencies. The early story of West Valley was a case in point.

An alternate way to market this service would be to treat nuclear fuel reprocessing as an environmental laundry, whose product is a clean environment, and whose customer is the general public. A laundry benefits by advertising its work to the public, and by competing openly with other providers of cleanliness. One way to realize this transition from confidential bargaining for cheapness to open competition for cleanliness might be to raise fuel reprocessing to the same rank with power production, and make it into a public service enterprise with a guarantee of fair returns on investment.

A clean environment is a good that the public is willing to buy. A competition to deliver this good would necessarily involve more than just the nuclear fuel reprocessing industry. It would pit processors of nuclear wastes against all other processors of industrial and municipal wastes, and atomic industries against other industries. As the public controversy over atomic energy broadens into a debate over all sources and uses of energy, there are signs that the AEC may enter into just this kind of competition. An openly publicized competition for cleanliness among all producers of energy could only benefit the public.

REFERENCES

Doub, W. O., 1972. Quality—the name of the game. *A.E.C. News Releases.* 3, no. 6:2–7.

Joint Committee on Atomic Energy, 1963. Hearing on the Nuclear Fuel Services, Inc. reprocessing plant in western New York State, May, 1963. *Atomic Energy Clearing House 9*, no. 20:12–31.

Klement, A. W., Jr., et al., 1972. *Estimates of ionizing radiation doses in the United States 1960–2000.* U.S.-E.P.A. Office of Radiation Programs, Rockville, Md.

Larson, C. E., 1972. International economic implications of the nuclear fuel cycle. *A.E.C. News Releases* 3, no. 32:3–11.

New York State. Department of Environmental Conservation, Albany, N.Y.

Environmental Radiation Bulletin, RAD-P1 (Quarterly).

Bentley, W. G., 1970. *Annual report of environmental radiation.*

———, 1971. *Annual report of environmental radiation.*

Thompson, M. H., (Dept. of Health), 1967. *Environmental radioactivity in New York State.*

———, 1968. *Environmental radioactivity in New York State.*

Nichols, J. P., and F. T. Binford, 1971. Status of noble gas removal and disposal. ORNL-TM-3515, Oak Ridge National Laboratory, Oak Ridge, Tenn.

Nuclear Fuel Services (NFS), independent surveys.

Barasch, G. E., and R. H. Beers, 1972. Aerial radiological area scanning surveys of the Nuclear Fuel Services plant, West Valley, New York, 1968 and 1969, ARMS-68.6.9. U.S. Atomic Energy Commission Technical Info. Ctr.

Cochran, J. A., et al. An investigation of airborne radioactive effluent from an operating nuclear fuel reprocessing plant, pp. 39. BRH/NERHL 70-3. U.S. Public Health Service Report.

Daly, J. C., et al., 1974. Iodine-129 levels in milk and water near a nuclear fuel reprocessing plant. *Health Physics* 26:333-342.

Davies, S., 1968. Environmental radiation surveillance at a nuclear fuel reprocessing plant. *Am. J. Public Health* 58:225k–2260.

Gillette, R., 1974. "Transient" nuclear workers: A special case for standards. *Science* 186:125–129.

Kelleher, W. J., 1969. Environmental surveillance around a nuclear fuel reprocessing installation. *Radiol. Health Data and Rep.* 10:329–339.

Lawroski, S., et al., 1963. Report of Study Group on the Nuclear Fuel Services Proposal, December, 1962, *At. Energy Clearing House* 9, no. 20:12–31.

Logsdon, I. E., and J. W. N. Hickey, 1971. Radioactive waste discharges to the environment from a nuclear fuel reprocessing plant. *Radiol. Health Data Rep.* 12:305–312.

Magno, P., et al., 1970. Liquid waste effluent from a nuclear fuel reprocessing plant. BRH-NERHL-70-2, National Techn. Info. Svce., Springfield, Va.

R.C.S.I. Industrial Radioactive Waste Bulletin No. 3, Feb. 28, 1968; No. 4, Sept. 17, 1968. Rochester Committee for Scientific Information, P.O. Box 5236, River Campus Station, Rochester, N.Y. 14627.

Sax, N.J., et al., 1969. Radioecological surveillance of the waterways around a nuclear fuels reprocessing plant. *Radiol. Health Data Rep.* 10:289–296.

Suda, S. S., 1970. Evaluation of I.A.E.A. safeguards inspection of Nuclear Fuel Services, West Valley, N.Y. BNL50228 (T-567), Brookhaven National Laboratory, Upton, Long Island.

Organization for Economic Cooperation and Development, 1972. Nuclear legislation. Analytical study. Regulations governing nuclear installations and radiation protection. Paris.

Orlans, H., 1967. *Contracting for Atoms.* The Brookings Institution, Washington, D.C.

Peterson, H. T., et al., 1969. Environmental tritium contamination from increasing utilization of nuclear energy sources. Bureau of Radiological Health (Environ. Control Admin., U.S. Dept. H.E.W.).

Sinclair, E. E., et al., 1971. Existing and projected plans and processes for thermal reactor fuel recovery: experience and plans. A/conf. 49/P-065, Session 2.4 pp. 1–12.

10 Code of Federal Regulations Part 50 Appendix F. Feb. 12, 1971.

U.S. Atomic Energy Commission Directorate of Licensing, Fuels and Materials. Environmental survey of the nuclear fuel cycle, Nov. 1972.

Wang, Y., ed., 1967. *Handbook of radioactive nuclides.* Cleveland: C.R.C. Co., pp. 573–836.

West Valley documents
indexed in: Nuclear Science Abstracts
available from:
Docket No. 50-201
U.S.A.E.C. Division of Technical Information
Oak Ridge, Tenn.

Nuclear Fuel Services, Inc., Rockville, Md., 1973. *Environmental report, NFS Reprocessing Plant, West Valley, New York.*

Nuclear Fuel Services, Inc., West Valley, N.Y., 1972. *Summary report No. 8. Low level waste treatment facility.*

Technical Specifications, Change No. 15, 1971. License CSF-1, Nuclear Fuel Services, Inc., and New York State Atomic and Space Development Authority.

Wenstrand, T. K., and D. P. Wilcox, 1972. Nuclear Fuel Services, Inc. *Environmental Report No. 11.*

Wilcox, D. P., and T. K. Wenstrand, 1971. Nuclear Fuel Services Inc. *Environmental Report No. 9.*

WILLIAM W. HAMBLETON

Storage of High Level Radioactive Waste[1]

The potential hazards of radioactive waste derive from the basic characteristics of the radioisotopic contaminants, although, in the words of a 1961 staff report of the Federal Radiation Commission, ". . . any radiation exposure involves some risk (to human health), the magnitude of which increases with the exposure . . ." Many radioisotopes decay rapidly; some decay at such a slow rate that they represent a potential hazard to mankind for centuries. Allowing these radioisotopes to decay naturally is the only practical means of reducing their radioactivity to non-hazardous levels. The isotopes that are of greatest concern are those which are highly toxic and have long lives, including strontium 90 and cesium 137, which require hundreds of years to decay, and plutonium 239, which has a half life of 24,000 years and requires more than 250,000 years to decay to an innocuous level—five times the history of Homo sapians on earth to date.

Radioactive wastes vary widely in the concentration of radioactive materials. High-level liquid wastes cannot be released into the environment because of their high radioactivity concentration, which may be as much as 10,000 curies per gallon. To confine and isolate high-level liquid wastes, the AEC has stored them underground in large steel-lined, concrete tanks and in steel tanks within concrete vaults. These liquid wastes from AEC operations, which now amount to some 80 million gallons, require continual surveillance, and storage in this manner can be considered only an interim solution.

Radioactive wastes containing numerous radioisotopic products have

[1]A revised form of this paper was published in the *Technology Review* (MIT), March/April, 1972, as "The Unsolved Problem of Nuclear Waste."

been generated in processing irradiated nuclear fuels at the chemical-processing plants operated by AEC's Richland, Savannah River, and Idaho Operations Offices, as well as at the commercial plant of Nuclear Fuel Services Incorporated, at West Valley, New York. Oak Ridge National Laboratory has generated high-level liquid wastes at its radiochemical-processing pilot plant, and is currently generating such wastes at its transuranium-processing facilities. Additional commercial fuel reprocessing plants are being, or will be, constructed to meet the requirements for processing increasing amounts of irradiated fuels which will be generated in nuclear powered electric plants. The waste from these plants will amount to an estimated 60 million gallons per year by the year 2000.

At present, Richland is proceeding with removal of strontium 90 and cesium 137 from high-heat liquid wastes and in-tank solidification of the remaining liquid. Removal of cesium and strontium enables the remaining wastes to decay to low-heat liquid within about five years, and Richland is currently developing a process to construct a facility for solidifying and encapsulating liquid strontium and cesium concentrates. Solidification of low-heat liquid wastes into salt cakes in tanks is considered to be an interim storage process until the acceptability of the process can be determined. In 1968 Richland was faced with a potentially serious situation with respect to the condition of its existing tanks, for some leaks had been detected.

Idaho National Reactor Testing Station is converting liquid waste to a granular, calcined material, which is stored in stainless-steel bins in underground concrete vaults, as an interim storage process. At Idaho, the burial grounds have been inundated on occasion by water from melting snow.

At Savannah River, wastes are segregated on the basis of their heat generation rates, and are immobilized in tanks by evaporation to salt crystals and sludges. A tank leak at Savannah River would be more serious than at Richland because the leakage would be expected to migrate into ground water. According to the AEC, burial practices followed by Richland, Savannah River, Idaho, and Oak Ridge have not resulted in releases of radioactivity beyond the confines of the burial grounds.

STORAGE CONCEPTS

Dupont and Company, the operator of the Savannah River plant, has proposed that radioactive, separation-process waste be permanently stored in caverns to be excavated in bedrock at the plant site. The concept suggests that storage for 100 million gallons of waste can be excavated in bedrock, approximately 1500 feet below grade, and consisting of six storage tunnels arranged in three pairs. After each storage tunnel is ready to receive waste, it is to be sealed from an access tunnel by impervious bulkheads, designed to withstand hydrostatic pressure at tunnel depth. The tunnels are to be constructed in predominantly Precambrian and Paleozoic metamorphic gneiss and schist, which are relatively impervious, but exhibit some fractures and fissures. These fracture and fissure zones can be sealed by grouting, and it is predicted that the migration of radioactive constituents from the storage facility will be so slow that no harmful contamination of off-site water will occur because of the low hydraulic gradient of bedrock water, the low solubility of plutonium, and the high density of the waste fluids as compared with groundwater. Geological, geochemical, hydrological, and economic aspects of the project have been investigated for almost eight years. A review panel of the National Academy of Sciences in May of 1969 concluded that the storage proposal has sufficient promise that construction of the shaft and several tunnels should be undertaken in order to determine the severity of the fracture problem. I note that in 1966 a majority of the Committee of the Earth Sciences Division of the National Academy of Sciences expressed strong reservation concerning the bedrock concept of waste storage, and recommended that the investigation be discontinued. More recently, a prominent member of that committee observed that the proposed storage at Savannah is a "disaster looking for a place to happen." The AEC has decided, however, to perform the additional studies because of the cost differential between bedrock storage and other alternatives.

Deep-cavern storage also has been proposed for Richland as an alternative to long-term storage of solidified waste in tanks. Studies were begun in 1969 to determine the feasibility of isolating wastes in caverns mined into basalt, 2000 to 4000 feet beneath the site. According to this concept, salt cake resulting from in-tank solidification of liquid waste would be removed from the tanks in the dry state, water would be added in the transfer system, and the slurry waste would be transported to the underground caverns. Richland is conducting a program of exploratory drilling to secure geological, hydrological, and other physical data to be used in evaluating the suitability of these subsurface rocks for waste storage.

Salt formations attracted the attention of a National Academy of Sciences Committee in 1955 because salt is abundant, can heal its own fractures by plastic flow, transmit heat readily, and exhibits compressive strength and radiation shielding properties similar to those of concrete. However, not until 1959 were studies relating to salt storage initiated at Oak Ridge. In 1963 studies were undertaken in Kansas, where the AEC chose a mine of the Carey Salt Company at Hutchinson for study of salt properties, and subsequently extended this study to the abandoned Carey Salt mine at Lyons. During Project Salt Vault studies at Lyons, engineering test-reactor fuel assemblies were utilized, along with heaters, to create an environment that would be similar to the expected environment of a real repository. The mine was instrumented with devices for recording heat, radiation, and physical properties.

Subsequently, the Lyons site was selected in 1970 as the actual storage location for radioactive wastes. This selection was based partly on research from the nearby Hutchinson mine and the Lyons mine. Other determining factors included the seismic stability of central Kansas, availability of a 300 foot section of salt overlain by 800 feet of rock containing impermeable shales, the generally flat-bedded character of the salt, the economic advantage of also using the abandoned Carey Salt mine for storage of low-level waste, and the hospitality of the people of Lyons.

Meanwhile, additional studies and design of storage facilities and methods to transport the radioactive waste were proceeding at Oak Ridge National Laboratory. According to plans, liquid wastes from commercial reprocessing plants would be converted into solid form and placed in stainless steel cylinders, which would be transported by rail in large shielded casks. Each cylinder would be lowered down a shaft into a newly excavated salt mine at Lyons. The cylinder would be placed in a hole in the floor of the mine with specially designed equipment. When an appropriate number of these cylinders had been placed in the mine, the entire room would be backfilled with crushed salt. Experimental evidence suggested that the crushed salt would recrystallize, and both crushed salt and bedded salt would flow plastically so as to completely seal-off the waste material. Other low-level radioactive materials consisting of contaminated clothing, lubricants, and laboratory ware were to be stored in the abandoned Carey Salt mine. There was some suggestion that granular, calcined waste from Idaho also eventually would be transferred to this repository.

The Kansas Geological Survey expressed serious concern about the proposal. One kind of concern related to the Lyons site itself; another related to the burial-in-salt concept. Initially, the geology of the area was inadequately known. On the basis of Survey recommendations, the AEC funded further geological studies. Numerous holes were drilled and logged at the site, water samples were taken in all the holes and analyzed, and a geological evaluation of the area has been conducted. These investigations revealed a probable major fault in the area and a pressure sink on the water surface of a major aquifer, suggesting vertical circulation. Furthermore, other conditions at Lyons were revealed that demonstrated the inadequacy of prior investigations. The abandoned Carey Salt mine is located on the north border of this town of approximately 5000 people at a depth of 800 feet. An entry in this mine extends southward beneath the City of Lyons. The only access to the mine is a vertical shaft, which penetrates 40 feet of saturated aquifer. The water is collected in a ring about 200 feet down the shaft, and pumped back to the surface. Just to the southwest of the underground mining operation, the American Salt Company also mines salt hydraulically by injecting fresh water, which dissolves the salt and creates caverns the full height of the salt. The resulting brine is returned to the surface for processing. The area contains both abandoned and producing oil and gas wells, numbering in the hundreds. The locations of some of these old wells have never been determined, and surface subsidence has occurred in places where old casing has corroded and permitted surface and groundwater to excavate caverns in the salt. Some of the resulting surface depressions are as much as 1/4 of a mile in diameter. Some of these wells penetrate deep Arbuckle rocks, which contain fluids under hydrostatic pressure sufficiently great that the static water level in the well stands higher than the level of salt mines. It is clear that intersection of an Arbuckle well by mining will cause flooding of the mine. At least 29 wells have been identified on the site to be acquired by the Atomic Energy Commission, and these must be completely cleaned out and replugged. We have reason to believe that other unidentified wells may be present. An abandoned shaft representing an earlier salt mining effort has been located just west of Lyons. This shaft is full of water.

Some time ago, the American Salt Company intersected an abandoned oil well with a rock bit, preparatory to shooting the salt face. Although some water entered the mine, the hole was plugged satisfactorily. Somewhat earlier, the American Salt Company lost all circulation during a hydraulic mining operation. Following successful injection of fresh water and production of brine for five days, approximately 180 thousand gallons of fresh water disappeared. The operation was terminated and no one can discover where the water went. In other words, the Lyons site is a bit like a piece of swiss cheese, and the possibility for entrance and circulation of fluids is great. Should fluids penetrate the American Salt Company mine, the possibility of salt solution and entrance into the Carey operation also is great.

Other investigations revealed approximately 400 feet of displacement in Arbuckle rocks, suggesting the presence of a major fault. All of these factors have led the Geological Survey to recommend that the Lyons site be abandoned. An independent analysis by a Committee of the Kansas Geological Society produced similar conclusions, as did an analysis by the Council of the Kansas Academy of Science. There is nothing more important than recognizing a dead horse early, and burying it with as little ceremony as possible.

As to the concept of burial in salt, the jury is still out. The axial temperature of the cylinders containing the radioactive waste is about 930°C. The cylinders will generate heat that must be dissipated through the salt and other overlying and underlying rocks. We claim that the two-layer, two-dimensional heat-flow model used by the AEC is overly simplified. A multi-layered, three-dimensional heat flow model is necessary for a clear description of the problem. This work is being undertaken at the present time at the Kansas Geological Survey. Another problem relates to possible mine subsidence. The crushed salt used to back-fill the mine will contain approximately 30% void space. Recrystallization and plastic flow of the salt could cause subsidence and shear in the overlying rocks, permitting surface or ground water to penetrate the mine, dissolve the salt, and set up a thermal transport system. This situation is even more dangerous because the stainless steel cylinders are expected to begin to break down within three months, releasing the waste. Likewise, the rock mechanical model used for studies of mine subsidence is overly simplified. Many rock properties are temperature dependent, and even dewatering of shales could create problems. Radiation damage and subsequent release of energy as a thermal excursion, both with respect to the salt and the radioactive waste itself, is an inadequately investigated subject. Appropriate studies should reveal whether or not radioactive waste can be stored safely in salt.

Meanwhile, the Kansas Geological Survey has undertaken additional reconnaissance studies of other areas of Kansas for possible storage of radioactive waste. The study is concerned with eight large areas that seemingly are underlain by salt beds that are at least 200 feet thick, no deeper than 2000 feet, and no shallower than 500 feet. Additionally, these areas contain a small number of oil and gas wells, salt mines, storage cavities, and pipelines, and a small population. A literature and file search has been undertaken for these areas to assemble information regarding salt and overburden thickness, quality of the salt, ground water conditions, and regional geological characteristics, as well as information about mineral resources, well locations, salt mines, liquid-petroleum-gas storage cavities, pipelines and population. The report will present an evaluation of these factors for each area. On the basis of these evaluations, the AEC will determine if any of the areas justifies further investigation. Because the areas contain few wells, information concerning the underlying rocks is sparse, and much additional investigation will be required before any of the areas can be judged suitable for storage of radioactive waste.

In July 1971, Cohen, Lewis, and Braun of Lawrence Radiation Laboratory proposed a method for disposing of nuclear reactor wastes by insitu incorporation in molten silicate rock. The proposal suggests injection of liquid wastes into a chimney formed by a 5 kiloton nuclear explosion at approximately a 2000 meter depth. The waste would be permitted to self-boil, and the resulting steam would be recycled and processed in a closed system. When waste addition is terminated, the chimney would be allowed to boil dry, thereby solidifying the wastes. The heat generated by the radioactive waste would then melt the surrounding rock, which would dissolve the waste. Finally the rock would refreeze, trapping the radioactive material underground in an insoluble rock matrix. The authors claim safe, permanent, and timely removal of radioactive material from the biosphere at relatively low cost, elimination of the need for waste transportation by siting in the immediate vicinity of the reprocessing plant, and waste injection with minimal or no treatment. Waste addition at a rate of 1500 tons per year for a period of 25 years is contemplated. The concept has been described as interesting and worthy of further study by numerous reviewers. However, it has been criticized on the grounds of conflict with the concept of recoverability, and serious doubts about the insolubility of the rock matrix, differentiation permitting plutonium to concentrate in a near-critical mass, geochemical alteration of the drill hole and casing, and gaseous phase transport of such compounds as ruthenium tetraoxide. Obviously, much research is necessary.

On September 16 and 17, 1971, Gisela Dreschoff and Edward J. Zeller, Research Associates of the Geological Survey of Kansas, visited the Asse Nuclear Waste Repository in Germany where key staff members provided a complete review of the project. At present, low-level waste is stored in a cavity in the Asse salt stock. Containers are released from shipping shields and lowered by crane into a chamber. Remote control facilities and remote television cameras permit movement and observation of the waste containers. No attempt is made to achieve symmetrical stacking. At the present time, nearly 10,000 casks of low-level waste, each having a limit of five curies total activity, are stored in two rooms at the 750 meter level. High-level waste emplacement is planned for 1974 or 1975. The waste will be solidified in the form of glass cylinders approximately 20 centimeters in diameter and one meter in length. These will be stacked vertically in bore holes in the salt roughly 50 meters deep in tunnels at the 750 meter level. After filling the bore holes with 30 meters of high-level waste cylinders, a concrete plug will be poured and the upper part of the hole above the plug will be filled with crushed salt. The Asse anticline is structurally stable and a massive gypsum cap rock can support the load of overlying sediments and serve as a shield for the underlying salt. Two nearby mines are flooded as a result of improper mining techniques. The Germans do not seem to be concerned, because the system has reached an equilibrium, no collapse has been observed near the old shafts, and no significant leakage has been determined. Even if the water entered the mine, the Germans feel the water would be nearly saturated and would cause no problems. However, the presence of sinkholes and salt springs does indicate that some solution is taking place.

Extensive studies are being conducted at the Hahn-Meitner-Institut for Nuclear Research in Berlin and at the Nuclear Research Center at Karlsruhe, which our colleagues also visited. These studies are concerned with radiation damage and the problems of processing and solidifying nuclear waste in form suitable for storage in the Asse mine. Competent scientists are in charge of the programs, which appear to be free of irrational political influence. Mention was made of the desirability of greater interchange of information between the United States and Germany regarding matters related to nuclear waste disposal. Seemingly, there has been little exchange in the past four years, and direct liaison between the U.S. and German research and development groups should be established as soon as possible. As far as can be determined, high-level radioactive waste is being stored at or near the surface in other countries.

At the recent Interstate Oil Compact meeting in Biloxi, Gene P. Morrell, Director of the Office of Oil and Gas, noted that as the decade of the 70s began, we are consuming energy at a rate of 68.8 quadrillion BTUs annually, and that requirements may reach 133 quadrillion BTUs by 1985. He concludes that our national energy policies have not kept pace with the rapid and unprecedented changes in our consumption patterns and social objectives. Superimposed upon this pattern, we find a developing nuclear power complex, that with its waste storage problem largely unresolved, will probably be called upon to provide up to 15% of our energy needs by 1985.

For the fiscal year 1970, AEC was authorized $2.3 billion for its various programs. Of this amount, only 28 million or roughly one percent represents operating and capital funds authorized for waste management programs. The Government Accounting Office believes that to expedite the development of methods for placing high-level waste in long-term isolation, AEC should place greater emphasis on evaluating the actions taken by its contractors, determining the adequacy of long-term storage proposals, and taking the steps needed to accomplish long-term storage. AEC has not established an overall, well-coordinated plan for resolving its waste management problems and achieving its objectives at all installations. In the past and currently, AEC management has emphasized and given priority to the development of technology and plans with respect to weapons production and reactor development which result in the generation of radioactive waste. Even where long-term storage has been of concern, the AEC has adopted the attitude that what is worth doing is worth doing wrong.

DONALD P. GEESAMAN

Plutonium and Public Health

On May 11, 1969, a major fire occurred at the large Rocky Flats plutonium facility located northwest of Denver, Colorado, and operated for the AEC by the Dow Chemical Company. For description of this fire see AEC press releases M-121, May 20, 1969, and M-257, November 18, 1969.

Consequent to this fire E. A. Martell and S. E. Poet conducted a pilot study on the plutonium contamination of surface soils in the Rocky Flats environs. Their results suggested an off site contamination that was orders of magnitude larger than that which would have been expected from the measured plutonium releases in the air effluent of the facility.

In a letter of January 13, 1970, to Glenn Seaborg, then chairman of the AEC, and in a press release of Febrauary 24, 1970, by the Colorado Committee on Environmental Information, Martell and others (1970) called attention to this anomalous contamination and expressed concern over its uncertain origin and over its significance to public health. In response the AEC fixed the probable origin of the off site contamination as wind dispersal of plutonium leaking from rusted barrels of contaminated cutting oil, and denied that cause existed for concern over hazards to public health (see AEC press release N-22, February 18, 1970).

It was my conviction that the AEC response provided a distorted and inadequate representation of the possible hazards assocated with the observed off site contamination, and that the imminent large-scale commercial introduction of plutonium gave this situation a precedential significance much greater than the already considerable significance of the situation itself.

In April 1970 a representative of the AEC's Division of Biology and Medicine and myself were invited to present our views at the University of Colorado. "Plutonium and Public Health" derives from the preceding history and should be so interpreted. The presentation was to a lay audience and was made with that expectation. Adequate referencing was added to the written text prior to its inclusion in *Underground Uses of Nuclear Energy, Part 2, Hearings before the Subcommittee on Air and Water Pollution of the Committee on Public Works* United States Senate, August 5, 1970.

As it stands the paper still represents a legitimate critique, and the recent emphasis on plutonium as a major energy source increases the relevance of the discussion. An updating would involve only incremental changes, and would generally supplement rather than disturb the substantive arguments of the original paper. Hence while such an updating is desirable, it is also of sufficiently marginal value that it can be properly deferred at my discretion.

For those who are interested in reading the traditional AEC position on the subject I would suggest "Appendix 24—Safety Considerations in the Operations of the Rocky Flats Plutonium Processing Plant," from *AEC Authorizing Legislation Fiscal Year 1971—Hearings before the Joint Committee on Atomic Energy, Part 4,* March 19, 1970.

Times have changed since May 1969. Then plutonium was regarded as a military substance and was accordingly given little public attention. Now it is much publicized as the energy source of the not too distant future. April 1970 was a time of transition, and I felt the strong presence of the earlier tradition, and the decision to speak was not an easy one for me. I have had no regrets.

—D. P. G.

PLUTONIUM AND PUBLIC HEALTH

For the sake of completeness let me give you some background on plutonium. It is an element that is virtually non-existent in the earth's natural crust. In the early 1940s it was first produced and isolated by Dr. Seaborg and colleagues. Plutonium has several isotopes, the most important being plutonium-239, which, because of its fissionable properties and its ease of production, is potentially the best of the three fission fuels. That is why it is of interest. Aside from its fissionable properties, plutonium-239 is a radioactive isotope of relatively long half-life (24,000 years), hence its radioactivity is undiminished within human time scales. When it decays, it emits a helium nucleus of substantial energy. Because of its physical characteristics, a helium nucleus interacts strongly with the material along its path; and as a consequence deposits its energy in a relatively short distance, about four-hundredths of a millimeter in solid tissue. For comparision, a typical cell dimension is about 1/4 to 1/10 of that. A cell whose nucleus is intercepted by the path of such a particle suffers sufficient injury that its capacity for cell division is usually lost (Bardenson, 1962; Bloom, 1959).

The cancer inducing potential of plutonium is well known. One millionth of a gram injected intradermally in mice has caused cancer (Lisco et al., 1947); a similar amount injected into the blood system of dogs has induced a substantial incidence of bone cancer (Mays et al., 1969), because of plutonium's tendency to seek bone tissue. Fortunately the body maintains a relatively effective barrier against the entry of plutonium into the blood system. Also, because of the short range of the emitted helium nuclei, the radiation from plutonium deposited on the surface of human skin does not usually reach any relevant tissue. Unfortunately the lung is more vulnerable.

Before I describe why this is, I'd like to say something about the characteristics of an aerosol. An aerosol is physically like cigarette smoke, or fog, or cement dust. Because of their small size, the particles comprising an aerosol remain suspended in air for long periods of time. If an aerosol is inhaled, then, depending on its physical characteristics, it may be deposited at different sites in the respiratory tree (*Health Physics,* 1966). Larger aerosol sizes are usually removed by turbulence in the nose; particles deposited in the bronchial tree are cleared upward in hours by the ciliated mucus blanket that covers the structure. This clearance system does not penetrate into the deep respiratory structures, the alveoli, where the basic oxygen-carbon dioxide exchange of the lung takes place. Smaller particles tend to be deposited here by gravitational settling, and if they are insoluble they may reside in the alveoli for a considerable time. The problem is that, under a number of conditions (Anderson and Nelson, 1967; Fraser, 1967; Kirchner, 1966; Mann and Kirchner, 1967; Stewart, 1963; Wilson and Terry, 1967) plutonium tends to form aerosols of a size that are preferentially deposited in deep lung tissue. Plutonium dioxide, which is a principal offender, is insoluble and may be immobilized in the lung for hundreds of days before being cleared to the throat or to the lymph nodes around the lungs (*Health Physics,* 1966).

An aerosol is comprised of particles of many different sizes, and their radioactivity may differ by factors of thousands or even more. I will simplify the argument and say that there is a class of these particles, the largest ones deposited in the deep lung tissue, that can be expected to have a different potential of cancer induction than the particles of the smaller class. This is because they are sufficiently radioactive to disrupt cell populations in the volume of cell tissue which they expose (Geesaman, 1968a). An example might be a particle that emits 5000 helium nuclei per day. It would subject between 1 and 20 alveoli to intense radiation, sufficient to inflict substantial cell death and tissue disruption. For reference, the alveoli are the basic structural units of the deep lung. They are shaped and bunched roughly like hollow grapes 0.3 millimeter in diameter. Their walls are thin, a few thousandths of a millimeter, and they are a highly structured tissue with many cell types. Intense exposure of local tissue by a radioactive particle is referred to as the hot particle problem. The question is: does such a particle have an enhanced potential for cancer? No one knows. One can argue that cancer cannot evolve from dead cells, hence a depleted cell population must be less carcinogenic. This is believable, and must be true on occasion. The facts are, though, that intense, local doses of radiation are extremely effective carcinogens, much more so than if the energy were averaged over a larger tissue mass (Geesaman, 1968b). Furthermore, this can take place at high doses of radiation where only one cell in ten thousand has retained its capacity to divide. The cancer susceptibility of lung tissue to radiation has been demonstrated in many species; one can say in general that the lung is more susceptible to inhomogeneous exposures from particles and implants than it is to diffuse uniform radiation. Some very careful skin experiments of Dr. Albert have indicated that tissue disruption is a very likely pathway of radioactive induction of cancer after intense exposure (Albert et al., 1967a, 1967b, 1967c, 1969). The experiments show that the most severe tissue injury is not necessary, nor even optimal, for the induction of cancer. When these notions are applied to a hot particle in the lung, the possibility of one cancer from 10,000 disruptive particles is realistic. This is disturbing because an appreciable portion of the total radioactivity in a plutonium aerosol is usually in the large particle component.

Let me demonstrate what I mean. Suppose a man received a maximum permissible lung burden for plutonium, and suppose roughly 10% of the mass of the burden was associated with the most active class of particles deposited (that is those emitting several thousand helium nuclei per day). This is reasonable. There would be something like a thousand of these particles and each would chronically expose 1 to 20 alveoli to intense radiation. If the risk of cancer is like 1 in 10,000 for one disruptive particle, then the total risk in this situation is one in ten, i.e., one man in ten would develop lung cancer.

Put another way, about 1 cubic centimeter of the lung is receiving high doses of radiation. It would not be surprising if intense exposure of such a localized volume led to a cancer one time in ten. The question is: if the individual volumes are separated from each other, is substantial protection afforded? No one knows. It is much easier to find two cancers using 50 exposures of 1 cubic centimeter each, than it is to find a couple of cancers in 50,000 single particle exposures. Certainly the length scales of injury are long enough that a disruptive carcinogenic pathway cannot be disregarded for isolated hot particles (Geesaman, 1968b).

One can look to the relevant experience for reassurance. In an experiment done at Hanford by Dr. Bair and his colleagues, beagle dogs were given $Pu^{239}O_2$ lung burdens of a few hundred thousandths of a gram (Bair et al., 1966; Ross, 1967). At 9 years post exposure, or after roughly half of an adult beagle life span, 22 of 24 deaths involved lung cancer, usually of multiple origin. Five dogs remain alive. For comparison, these exposures are about 100 times larger than the present maximum permissible burdens in man.

There are two unsatisfactory aspects of this experiment. First, because all of the dogs are developing cancer, it is impossible to infer what would happen at lower exposures; simple proportionality does, however, suggest that present human standards are too lax by at least a factor of ten. Second, because the radiation dose is large, with tissue injury almost killing the dogs; and because large numbers of particles are involved, often acting in conjunction; it is improbable that the risk from disruptive particles can be inferred. And after all, this is what we need to know, since almost all human exposures will involve hot particles acting independently, and if there is a risk from these particles, it will be additive throughout the population—there will be no question of a threshold burden; and there will be a possibility that a man with an undetectable burden of a few particles will develop a cancer as a consequence. For the exposures of concern, 1000 people with 100 disruptive particles each will suffer as many total cancers as 10,000 people with 10 particles each, or as 100 people with 1000 particles each.

Human experience does not give us the answer either. Plutonium has been around for 25 years, and people have been exposed. In 1964 through 1966, contractors indicated an average total of 21 people per year with over 25% of a maximum permissible burden of plutonium (Ross, 1968). Three out of four of these exposures derived from inhalation. To be reasonably useful, the documentation of exposure must go back more than 15 years, because of the latent period for radiation induced cancer. In recent years documentation has improved greatly, but from early days there is pitifully little of relevance to the hot particle problem in the lung.

Since I have mentioned maximum permissible lung burdens, you are aware that there is official guidance. I would like to comment on it. The maximum permissible lung burden is established by equilibrating the exposure from the deposited radioactive aerosol with that of an acceptable uniform dose of X rays. The International Commission on Radiological Protection indicates this may be greatly in error, and specifically states in its publication 9, "In the meantime there is no clear evidence to show whether, with a given mean absorbed dose, the biological risk associated with a non-homogeneous distribution is greater or less than the risk resulting from a more diffuse distribution of that dose in the lung." (ICRP, 1966). They are effectively saying that there is no guidance as to the risk for non-homogeneous exposure in the lung, hence the maximum permissible lung burden is meaningless for plutonium particles, as are the maximum permissible air concentrations which derive from it.

So there is a hot particle problem with plutonium in the lung, and the hot particle problem is not understood, and there is no guidance as to the risk. I don't think there is any controversy about that. Let me quote from Dr. K. Z. Morgan's testimony in January of this year before the Joint Committee on Atomic Energy, U.S. Congress (Morgan, 1970. Dr. K. Z. Morgan is one of the United States' two members to the main Committee of the International Commission on Radiological Protection; he has been a member of the committee longer than anyone; and he is director of Health Physics Division at Oak Ridge National Laboratory. Dr. Morgan testified, "There are many things about radiation exposure we do not understand, and there will continue to be uncertainties until health physics can provide a coherent theory of radiation damage. This is why some of the basic research studies of the USAEC are so important. D. P. Geesaman and Tamplin have pointed out recently the problems of plutonium-239 particles and the uncertainty of the risk to a man who carries such a particle of high specific activity in his lungs." At the same hearing, in response to the committee's inquiry about priorities in basic research on the biological effects of radiation, Dr. M. Eisenbud, then Director of the New York City Environmental Protection Administration, in part replied, "For some reason or other the particle problem has not come upon us in quite a little while, but it probably will one of these days. We are not much further along on the basic question of whether a given amount of energy delivered to a progressively smaller and smaller volume of tissue is better or worse for the recipient. This is another way of asking the question of how you calculate the dose when you inhale a single particle." (Eisenbud, 1970). He was correct; the problem has come up again.

In the context of his comment it is interesting to refer to the National Academy of Sciences–National Research Council report of 1961 on the Effects of Inhaled Radioactive Particles (NAS–NRC, 1961). The first sentence reads, "The potential hazard due to airborne radioactive particulates is probably the least understood of the hazards associated with atomic weapons tests, production of radioelements, and the expanding use of nuclear energy for power production." A decade later that statement is still valid. Finally let me quote Drs. Sanders, Thompson, and Bair from a paper given by them (Sanders et al., 1970). Dr. Bair and his colleagues have done the most relevant plutonium oxide inhalation experiments. "Nonuniform irradiation of the lung from deposited radioactive particulates is clearly more carcinogenic than uniform exposure (on a total-lung dose basis), and alpha-irradiation is more carcinogenic than beta-irradiation. The doses required for a substantial tumor incidence, are very high, however, if measured in proximity to the particle; and, again, there are no data to establish the low-incidence end of a dose-effect curve. And there is no general theory, or data on which to base a theory, which would permit extrapolation of the high incidence portion of the curve into the low incidence region." I agree and I suggest that in such a circumstance it is appropriate to view the standards with extreme caution.

There is another hazardous aspect of the particulate problem in which substantial uncertainty exists. In case of an aerosol depositing on a surface, the material may be resuspended in the air. This process is crudely described by a quantity called a resuspension factor which is remarkable in that it seems generally known only to within a factor of billions (Kathren, 1968). Undoubtedly it can be pinpointed somewhat better than this for plutonium oxide, but the handiest way to dispatch the problem is to say there is some evidence that plutonium particles become attached to larger particles and are therefore no longer potential aerosols. Unfortunately there is also evidence that large particles generate aerodynamic turbulence, and are hence blown about more readily, and on being redeposited tend to knock small particles free. In relation to this, I'd like to give you a little subjective feeling for the hazard. There is no official guidance on surface contamination by plutonium. Two years ago, in an effort to determine some indication of the opinions of knowledgeable persons with respect to environmental contamination by plutonium, a brief questionaire was administered to 38 selected LRL employees (Kathren, pers. comm.). All were persons who were well acquainted with the hazards of plutonium. The group consisted of 16 Hazards Control personnel, primarily health physicists and senior radiation monitors. The remainder were professional personnel from Biomedical Division, Chemistry, and Military Applications, who had extensive experience with plutonium. I had nothing to do with the survey, nor was I one of the members who was queried. The conjectured situation was that their neighborhood had been contaminated by plutonium oxide to levels of 0.4 microcuries per square meter. For reference, this value is roughly ten times the highest concentration Dr. Martell found east of the Rocky Flats Dow Chemical facility (Martell et al., 1970), and bear in mind that a factor of ten is a small difference relative to the large uncertainties associated with the hazards from plutonium contamination. Several questions were asked. Would you allow your children to play in it? 86% said No. Should these levels be decontaminated? 89% said Yes. And as to the level the area should be cleaned, 50% said to background, zero, minimum, or by a reduction of at least a factor of 40. This has no profound scientific significance, but indicates that many people conversant of the hazard are not blasé about the levels of contamination encountered east of Rocky Flats.

Finally I would like to describe the problem in a larger context. By the year 2000, plutonium-239 has been conjectured to be a major energy source. Commercial production is projected at 30 tons per year by 1980, in excess of 100 tons per year by 2000. Plutonium contamination is not an academic question. Unless fusion reactor feasibility is demonstrated in the near future, the commitment will be made to liquid metal fast breeder reactors fueled by plutonium. Since fusion reactors are presently speculative, the decision for liquid metal fast breeders should be anticipated and plutonium should be considered as a major pollutant of remarkable toxicity and persistence. Considering the enormous economic inertia involved in the commitment it is imperative that public health aspects be carefully and honestly defined prior to active promotion of the industry. To live sanely with plutonium one must appreciate the potential magnitude of the risk, and be able to monitor against all significant hazards.

An indeterminate amount of plutonium has gone off site at a major facility 10 miles upwind from a metropolitan area. The loss was unnoticed. The origin is somewhat speculative as is the ultimate deposition.

The health and safety of public and workers are protected by a set of standards for plutonium acknowledged to be meaningless.

Such things make a travesty of public health, and raise serious questions about a hurried acceptance of nuclear energy.

REFERENCES

Albert, R. E., et al., 1967a. The effect of penetration depth of electron radiation on skin tumor formation in the rat. *Radiation Res.* 30:515–524.

———, 1967b. Skin damage and tumor formation from grid and sieve patterns of electron and beta radiation in the rat. *Radiation Res.* 30:525–540.

———, 1967c. The association between chronic radiation damage of the hair follicles and tumor formation in the rat. *Radiation Res.* 30:590–599.

———, 1969. An evaluation by alpha-particle Bragg peak radiation of the critical depth in the rat skin for tumor induction. *Radiation Res.* 39:332–344.

Anderson, B. V., and I. C. Nelson, 1967. Plutonium air concentrations and particle size relationship in Hanford facilities. BNWL-495, Dec. 1967.

Bair, W. J., et al., 1966. Long-term study of inhaled plutonium in dogs. Battelle Memorial Institute Technical Report, AFWL-TR-65-214.

Barendson, G. W., 1962. Dose-survival curves of human cells in tissue culture irradiated with alpha-, beta-, 20-kV x- and 200-kV x-radiation. *Nature* 193:1153–1155.

Bloom, W., 1959. Cellular responses. *Rev. Mod. Phys.* 31:21–29.

Eisenbud, M. Panel discussion, 1970. In *Environmental effects of producing electrical power, phase 2.* Testimony presented at Hearings before the Joint Committee on Atomic Energy, 91st Cong., 1970. Washington, D.C., U.S. Gov't. Print. Off, forthcoming.

Fraser, D.C., 1967. Health physics problems associated with the production of experimental reactor fuels containing PuO_2. *Health Phys.* 13:1133–1143.

Geesaman, D. P., 1968a. An analysis of the carcinogenic risk from an insoluble alpha-emitting aerosol deposited in deep respiratory tissue. University of California Radiation Laboratory, Livermore, UCRL-50387.

———, 1968b. An analysis of the carcinogenic risk from an insoluble alpha-emitting aerosol deposited in deep respiratory tissue; Addendum, University of California Radiation Laboratory, Livermore, UCRL-50387, Addendum.

Health Phys., 1966. Task Group—Chairman, Paul E. Morrow, Deposition and retention models for internal dosimetry of the human respiratory tract, 12:173–207.

ICRP, 1966. *Recommendations of the International Commission on Radiological Protection (Adopted September 17, 1965), ICRP Publication 9.* Oxford: Pergamon Press. ICRP-PUBL-9.

Kathren, R. L., 1968. Towards interim acceptable surface contamination levels for environmental PuO_2. BNWL-SA-1510.

———. Battelle Northwest, personal communication.

Kirchner, R. A., 1966. A plutonium particle size study in production areas at Rocky Flats. *Am. Ind. Hyg. Assoc. J.* 27:396–401.

Lisco, H., et al., 1947. Carcinogenic properties of radioactive fission products and of plutonium. *Radiology* 49:361–363.

Mann, J. R., and R. A. Kirchner, 1967. Evaluation of lung burden following acute inhalation exposure to highly insoluble PuO_2. *Health Phys.* 13:877–882.

Martell, E. A., et al., 1970. Report on the Dow Rocky Flats fire: Implications of plutonium releases to the public health and safety. Colorado Committee for Environmental Information, Subcommittee on Rocky Flats, Boulder, Colorado, January 13, 1970. (Personal communication to Dr. Glenn T. Seaborg, Chairman, Atomic Energy Commission.)

Mays, C. W., et al., 1969. Radiation-induced bone cancer in beagles. In *Delayed effects of bone-seeking radionuclides,* ed. C. W. Mays. Salt Lake City: Univ. of Utah Press.

Morgan, K. Z. Radiation standards for reactor siting. In *Environmental effects of producing electrical power, phase 2.* Testimony presented at Hearings before the Joint Committee on Atomic Energy, 91st Cong., 1970. Washington, D.C., U.S. Gov't. Print. Off, forthcoming.

NAS-NRC-SUBCOMM, 1961. *Effects of inhaled radioactive particles.* Report of the Subcommittee on Inhalation Hazards. Committee on Pathologic Effects of Atomic Radiation. National Academy of Sciences-National Research Council, Washington, D.C., Publication 848. NAS-NRC/PUB-848, 1961.

Park, J. F., et al., 1970. Chronic effects of inhaled $^{239}PuO_2$ in beagles. BNWL-1050, Part 1:3.3–3.5.

Ross, D. M., 1968. A statistical summary of United States Atomic Energy Commission contractors' internal exposure experience, 1957–1966. In *Diagnosis and treatment of deposited radionuclides,* eds. H. A. Kornberg and W. D. Norwood. Proceedings of a Symposium held at Richland, Washington, 15–17 May 1967. N.Y., Excerpta Medica Foundation, 1968. pp. 427–434. (CONF-670521).

Sanders, C. L., et al., 1970. Lung cancer: Dose response studies with radionuclides. In *Inhalation carcinogensis.* Proceedings of a Biology Division, Oak Ridge National Laboratory, conference held in Gatlinburg, Tenn., October 8–11, 1969. M. G. Hanna, Jr., P. Nettesheim, and J. R. Gilbert, (eds.) U.S. Atomic Energy Commission Symposium Series 18, 1970. pp. 285–303, (CONF-691001).

Stewart, K., 1963. The particulate material formed by the oxidation of plutonium. In *Technology, engineering and safety,* ed. C. Nichols. New York: Macmillan 5:535-579.

Wilson, R. H., and J. L. Terry, 1967. Biological studies associated with a field release of plutonium. In *Inhaled particles and vapours II,* ed. C. Davies. Oxford: Pergamon Press, pp. 273–290.

JOHN W. GOFMAN AND ARTHUR R. TAMPLIN

Radiation: The Invisible Casualties

The Federal Radiation Council establishes the limits on nuclear radiation the U.S. population may be exposed to from peaceful atomic-energy activities. Most people assume that these standards, set by a responsible body, have a wide margin of safety built into the permissible dosage. Unfortunately, this simply is not true.

If the population of this country did receive the present maximum allowable radiation dosage, there would be an increase in the frequency of cancer and leukemia. The increase, conservatively speaking, would eventually be on the order of 24,000 additional cases of leukemia and cancer each year.

This figure is found by estimating the risk of various forms of cancer and leukemia per unit of radiation exposure and then calculating the effect if the entire population received the maximum allowable exposure. All forms of cancer show similar increases in incidence per unit of exposure for adults. This increase is approximately one percent in incidence of cancer or leukemia per year per Rad of exposure. (One Rad—"radiation absorbed dose"—is the standard unit of measurement of radiation exposure. Radiation exposure from natural sources averages about 0.15 Rad.)

The allowable dose is 0.17 Rad per person per year for the population at large. If everyone in the population received this dose from birth to age thirty years, the total exposure (the dose accumulated through the years over and above natural background radiation) would be 5 Rads per person. Following the evidence that the risk for all forms of cancer plus leukemia, is an increase of 1% per Rad, we estimate a 5% increase in all forms of cancer plus leukemia, if the population receives the maximum allowable exposure.

There are 200 million persons in the U.S., and roughly one-half are over thirty years of age. Cancers may appear many years after first exposure. If the general environmental levels of radioactivity are increased to the maximum allowable extent, the entire population will receive additional radiation exposure from the moment of conception, but not all of the physical disease caused by this exposure will appear until the exposed persons have reached age thirty. The spontaneous cancer incidence for people over thirty is approximately 280 per 100,000 persons per year. A 5% increase would mean that 14 additional cancer and leukemia cases per 100,000 persons per year should be expected due to radiation. For the entire over-30 population, this equals 14,000 additional cancer plus leukemia cases per year in the U.S., if the size, makeup, and cancer incidence of the general population does not change. Long lag-times plus lower accumulated dosage means a smaller number of additional cases would appear in the under-30 age group. It would by no means be an overestimate to add 2000 additional cases for this part of the population. This may be an underestimate, considering the greater sensitivity of this group to radiation-induced cancer. If the U.S. population does not grow, and other factors affecting the induction of cancer do not change, a total of 16,000 additional cases per year would be expected after 30 years. If the population continues to grow, as indeed it will, the number of people exposed and the number of cancers and leukemias will also grow. If the population increases by 50% in the next 30 years, not an unreasonable assumption, there will be 24,000 additional cancer and leukemia cases caused by radiation each year. This estimate does not take into account that 0.13 Rad will have been received by each infant before birth (0.17 Rad/year × 40/52 years). It is difficult to know whether this in-utero radiation carries an increased cancer risk for the whole lifetime or not. The additional contribution for the in-utero radiation (at a period when the effectiveness per Rad is very high) could be between a few hundred and several thousand additional cancer cases per year. We shall not attempt to guess the precise additional contribution to in-utero irradiation.

The Federal Radiation Council, which sets the allowable peacetime radiation standards, has access to the information by which we have arrived at the above conclusions. The present standard, if allowed to stand, must be understood to mean that the government is willing to trade off thousands of cases of cancer and leukemia in return for peaceful atomic-energy activities.

The above argument does not mean to imply that the population of the United States is currently receiving the maximum allowable dosage per person. We are far from approaching that level today. But it must be noted that the peaceful atomic-energy industry is just beginning. We may expect to see rapid growth within this decade which will add significantly to the population dosage. By 1980, approximately 200 nuclear power stations are expected to generate one-third of the nation's electricity; peaceful atomic explosives may be used to produce natural gas, mine copper, and dig harbors. Although industry does not plan to deliver the allowable dose, it would be legally possible to approach the standard. Now, when the atomic industry is first beginning, it would surely be easier and more economical in the long run for industry to meet a real standard (in place of an inflated one) that can be expected to hold up over time. A real standard now will avoid the distinct possibility of a later necessary revision downward of the standard. Such a revision would lead to very expensive retrofits in developed, industrial applications of nuclear energy. Retrofits will not be necessary if tooling-up is done correctly, with maximum safety margins, the first time.

Since the standard was established, few have questioned its justification. Indeed, there is very little justification for the standard—merely that at the set level, risk is *believed* to be small compared with the benefits to be derived from the orderly development of atomic energy for peaceful purposes. Yet some have questioned the validity of this argument. Dr. Brian MacMahon, Professor of Epidemiology at Harvard, writing as recently as early 1969, stated:

> While a great deal more is known now than was known 20 years ago, it must be admitted that we still do not have most of the data that would be required for an informed judgment on the maximum limits of exposure advisable for individuals or populations. (MacMahon, 1969).

In general we agree with Dr. MacMahon. However, we believe the evidence available indicates grave dangers to the population if the set level is ever attained. Two manifestations of these dangers, as stated above, are leukemia and cancer. There are surely other dangers—potentially even more serious—concerning genetic defects, fetal deaths, and neonatal deaths (Tamplin, 1969). We, however, limit ourselves here to discussion of leukemia and various forms of cancer in humans.

Much of the complacency that is prevalent today grew out of the time when there was a great paucity of data concerning the dose-versus-effect relationship between radiation and cancer or leukemia induction in man. We still can only estimate dose-effect rates for low doses of radiation. Unfortunately, most of the hard data concerning dose-effect relationships in man are for total doses of above 100 Rads. Our estimates, therefore, of the effect per Rad are, to be conservative, based upon a linear extrapolation from high dosages down to very low dosages. We assign the same cancer-producing risk to each unit of radiation, whatever the total dose of radiation may be.

There is some evidence that this assumption may, in the future, be proved incorrect. Stewart, Kneale, and MacMahon have worked on the effects of low doses of radiation (Stewart et al., 1958; Stewart and Kneale, 1968; MacMahon, 1962; MacMahon and Hutchinson, 1964). Their work, however, is on infants in-utero; this group of the population is most susceptible to radiation. The data did show that at two or three Rads the effect per Rad on infants in-utero may be even ten times more effective in inducing cancer than is each Rad at higher doses. On the other hand, it is possible, although doubtful, that at extremely low doses, ionizing radiation has no effect on man. The dose below which no harm occurs to man (if this dose does indeed exist) would be called the "threshold" dose. We and others are actively doing experiments on human cells to determine the effect per Rad at various total doses to see if threshold effects ever exist for man. However, it would be presumptuous, at best, to assume that the dose-effect rate is not linear from high to low or that a threshold dose does exist when we are considering the safety of present radiation guidelines.

Nor should we assume, at this time, that it is less harmful for man to receive radiation over a period of years, as would be the case for peaceful applications of atomic energy, than if the same dose were delivered rapidly. It is true that for some biological effects the ability of the body to repair damage from previous radiation makes the effect of slow, protracted radiation less harmful than for the same dose delivered rapidly; however, no evidence exists for such an effect on cancer or leukemia induction by radiation in man. Furthermore, uranium miners receive their radiation slowly, and it appears that any protection this provides, if there is any, is not sufficient to appreciably alter any of our major considerations. These assumptions, we should add, are shared by the Federal Radiation Council, and explicitly set out in their reports.

SOURCES OF INFORMATION

Several important sources have shed light on cause-and-effect relationships between irradiation and cancer induction in man. These sources include the following: (a) Study of survivors of Hiroshima and Nagasaki by the Atomic Bomb Casualty Commission; (b) Study of patients treated for nonmalignant diseases, who then developed cancer or leukemia later in life; (c) Study of children who, during an unfortunate era in American medicine, received irradiation for treatment of enlarged thymus and then developed thyroid cancer; (d) Study of the occurrence of lung cancer in uranium miners in the U.S.; (e) Study of cancer and leukemia in children whose mothers had been given diagnostic X rays during pregnancy.

One of the earliest sources of informaton on cause-and-effect relationships between cancer and irradiation was the Atomic Bomb Casualty Commission (ABCC). As this commission began reporting results of studies, it was noted that those persons irradiated in Hiroshima and Nagasaki might have a higher leukemia rate than normal. From the ABCC studies (Maki et al., 1968), and from wholly independent observations, it is now clear, that at least for total doses of 100 Rads or more, the leukemia risk may be expressed as follows: one to two cases of leukemia per million exposed persons per year, where each of them has received one Rad of total body exposure. This does not require one Rad per year. Rather, we are discussing the above rate of disease occurrence for a total accumulated exposure of one Rad. Furthermore, this incidence of one to two cases per million people per year persists for many years once the latency period is over, ultimately declining somewhat, at least for chronic leukemia.

An incidence rate of one or two cases per million people per year seems like a small number, especially when viewed in isolation. Indeed, many have hastened to point out that leukemia occurs spontaneously, without any radiation induction, at a frequency of 60 cases per million per year. Even at that rate leukemia is considered a relatively rare disease. Thus, one or two cases per million people seemed at first to be insignificant. What is more, leukemia was originally thought to be a "special" response to ionizing radiation. Not much thought was given to other forms of cancer. Therefore, thinking that leukemia was the only response to radiation, and that that response was insignificant, wide-spread complacency concerning long-term effects of radiation set in—a complacency extending to high circles.

For two reasons this complacency has turned out to be a dangerous mistake. Since the first ABCC reports it has been learned that leukemia happens to show a shorter latency period than most other forms of cancer. Therefore, the reason it appeared early in the studies to be the only malignancy in the Hiroshima-Nagasaki survivors was simply that not enough time had elapsed for the other forms of cancer to manifest themselves. Secondly, the proper way to look at the leukemia incidence rate of one to two cases per million persons per year from radiation and the 60 per million persons per year spontaneously is not in isolation from each other, but in relation to each other. Viewed in this light, one Rad increases the leukemia incidence between 1.6 and 3.3 percent. Or, we can state that the doubling dose for leukemia (namely, that amount of radiation which will double the spontaneous rate) is between 30 and 60 Rads. Doubling a spontaneous rate of 60 cases per million each year means producing an additional 60 cases per million per year.

It now has become clear that other forms of cancer respond to ionizing radiation in a manner similar to leukemia. That is, other forms of cancer are, like leukemia, describable by a fractional increase in occurrence rate per Rad. Let us now consider a variety of forms of human cancer other than leukemia and the effect of radiation upon any increased incidence rate. The supporting data for the arguments below are available from worldwide data, U.S. data, and from studies by the Atomic Bomb Casualty Commission of survivors of Hiroshima and Nagasaki.

Thyroid Cancer

The Japanese data, primarily based upon adults, show an approximate doubling dose of 100 Rads for development of thyroid cancer, or approximately a 1% increase in incidence rate of thyroid cancer in the population per Rad of exposure of the population (Maki et al., 1968).

We can arrive at the risk for younger people in the U.S. from two items of data: (a) Pochin gives the figure of one case of thyroid cancer per million persons per Rad (Pochin, 1966); (b) Carroll and others reported that the spontaneous incidence rate for thyroid cancer is approximately five to ten cases per million persons per year in the age range of 10 to 20 years (Carroll et al., 1964).

Combining these two items of information, it is estimated that between 5 and 10 Rads is the doubling dose for thyroid cancer in young people in the United States. This means a 10 to 20% increase in risk of thyroid cancer in the youthful population per year per Rad of exposure. Thus, considering the youthful group (U.S.) and the adults (Japan), the range is between 1% and 20% increase in thyroid cancer per year per Rad of exposure.

Lung Cancer

Estimates are available from several sources for radiation induction of lung cancer. The ABCC studies in Japan indicate an approximate doubling of lung cancer incidence rate for 100 Rads of exposure, or a

1% increase in risk of lung cancer in the population for an exposure of one Rad. (Maki et al., 1968). The experiences of the uranium miners in the U.S. are complicated by two factors (JCAE, 1967): (a) the actual exposures are not accurately known, and (b) many of the workers are still in the latency period. What estimates have been made for the uranium miners suggest the doubling dose for lung cancer to be between 250 and 500 Rads. If the correction for latency is estimated as twofold the final estimate would be 125 to 250 Rads as the doubling dose (JCAE, 1967).

Miller has questioned the Japanese data because of nonspecificity of the histology of the cancer cells (Miller, 1969). On the other hand, the similarity of the ratio of lung cancer to leukemia in the Japanese as compared to the British patients studied by Court-Brown and Doll suggests the Japanese data to be quite reasonable (Court-Brown and Doll, 1965). As a compromise estimate, we shall average the Japanese and U.S. data, to obtain 175 Rads as the estimate for the doubling dose for lung cancer, or a 0.6% increase in the annual incidence rate of lung cancer in the population per Rad of exposure.

Breast Cancer and Other Forms

Breast cancer has been found to be radiation-induced in the Japanese studies. The estimated doubling dose is approximately 100 Rads for breast cancer, or, again, a one percent increase in incidence rate per year of breast cancer in the population per Rad of exposure.

From some important studies on humans receiving therapeutic radiation for the arthritis-like disorder known as rheumatoid spondylitis, Court-Brown and Doll have studied the subsequent occurrence of many forms of cancer in organs heavily exposed, incidental to irradiation of the primary disease in the spine. We don't know that all the heavily exposed regions received equivalent doses, but it appears reasonable to estimate that the various heavily exposed regions were within a factor of 2 on either side of the median value of the group. If we use Court-Brown and Doll's value for bronchiogenic cancer of the lung as a reference value (and for this form of cancer we have used 175 Rads above as an estimated doubling dose), we can then estimate the doubling dose for radiation for several additional cancers. Uncertainty of precise dose comparisons makes these numbers uncertain by a factor of two or thereabouts either on the low or high side. We shall, therefore, not only show the estimated doubling doses for all these additional cancers, but also a range to take this dose uncertainty into consideration. Thus, we have for the following additional cancers:

Site of Cancer	Doubling Dose (Rads)		% Increase in Incidence Rate per Rad	
	Mean	Range	Mean	Range
Pharynx	40 Rads	(20–80)	2.5%	(1.2–5.0)
Stomach	230 Rads	(115–460)	0.4%	(0.2–0.8)
Pancreas	125 Rads	(60–250)	0.8%	(0.4–1.6)
Bone[a]	40 Rads	(20–80)	2.5%	(1.2–5.0)
Lymphatic plus other hematopoietic organs	70 Rads	(35–140)	1.4%	(0.7–2.8)
Carcinomatosis of miscellaneous origin	60 Rads	(30–120)	1.7%	(0.9–3.4)

[a]Bone may possibly have received a higher radiation does than other sites. If this was true, the estimated doubling dose is too low for bone.

This table summarizes the radiation-induced cancers for all sites, utilizing all the data available.

Best Estimates of Doubling Dose of Radiation for Human Cancers and the Increase in Incidence Rate per Rad of Exposure

Organ Site	Approximate Doubling Dose	% Increase in Incidence Rate per Rad
Leukemia	30–60 Rads	1.6–3.3%
Thyroid Cancer		
(adults)	100 Rads	1%
(young persons)	(5–10 Rads)	(10–20%)
Lung Cancer	175 Rads	0.6%
Breast Cancer	100 Rads	1%
Stomach Cancer	230 Rads	0.4%
Pancreas Cancer	125 Rads	0.8%
Bone Cancer	40 Rads	2.5%
Lymphatic & Other Hematopoietic organs	70 Rads	1.4%
Carcinomatosis of miscellaneous origin	60 Rads	1.7%

For such an array of widely divergent organ systems, already including hard data for nearly all the major forms of human cancers, it is amazing indeed that there is such a small range for the estimated doubling dose. Correspondingly, there is a very small range in the estimated increase in incidence rate per Rad for these widely differing organ sites in which cancers arise.

The only number that is different, and that one indicates an even higher susceptibility to radiation induction of cancer, is for thyroid-cancer induction in youthful persons (under twenty years of age). As we shall see, this is not at all surprising or inconsistent, for the data presented in the following table suggest a very high sensitivity of embryos in-utero to irradiation, causing subsequent leukemia and cancer during early childhood.

Furthermore, in some of these studies, aside from leukemia, the persons at risk were most probably still in the latency period when studied, so that full expression of the disease has not yet been reached. This would mean that an even smaller radiation dose is required to double the incidence rate. We know, from extensive other data, that bone cancer and skin cancer have definitely been produced by radiation. With further observation and study, the ABCC data will provide firm estimates of the doubling dose for the induction of cancer by radiation at the few remaining other major organ sites. At present, the only malignant disease reputedly not induced by radiation is chronic lymphatic leukemia. And even this may be in doubt, since malignant lymphoma, a highly related cancerous disorder, is radiation induced, both from the data of Court-Brown and Doll and from Japanese data (Anderson and Ishida, 1964).

Development of Childhood Leukemia and Cancer

Stewart and co-workers originally and MacMahon and Stewart and Kneale recently have presented evidence that implicates in-utero radiation of embryos (carried out for diagnostic purposes in the mother) with the development of subsequent leukemia plus other cancers in the first ten years of life of the child. The general estimate of the amount of radiation delivered in such diagnostic procedures is 2 to 3 Rads to the developing fetus. From the Stewart and Kneale data, we have, for the following forms of cancer, the estimates of the increase in numbers of cancers for several organ sites:

Type of Cancer	Radiation Induced Increase
Leukemia	50% increase over spontaneous incidence
Lymphosarcoma	50% " " " "
Cerebral Tumors	50% " " " "
Neuroblastoma	50% " " " "
Wilms' Tumor	60% " " " "
Other cancers	50% " " " "

From the MacMahon data, we have the following highly similar estimates:

Leukemia	50%	increase	over	spontaneous	incidence
Central Nervous System Tumors	60%	"	"	"	"
Other Cancers	40%	"	"	"	"

From both the MacMahon evidence and the Stewart-Kneale evidence, we have as a best estimate a 50% increase in incidence rate for all forms of cancer plus leukemia associated with diagnostic irradiation of the infant-in-utero; the numbers are closely similar for U.S. practice and British practice. So, for 2 to 3 Rads to the infant in-utero, a 50% increase in incidence rate of various cancers leads to an estimate of 4 to 6 Rads as the doubling dose for childhood leukemia plus cancer due to diagnostic irradiation in-utero. Let us underestimate the risk, and use the higher number, 6 Rads, as the doubling dose for in-utero induction of subsequent leukemia plus other childhood cancers. This means a 17% increase in the incidence rate of such leukemia plus cancers per Rad of in-utero exposure of the infant.

It is not at all surprising that infants in-utero should appear most sensitive to irradiation, children next in sensitivity, and adults third (but by no means low). This is precisely the order in which these groups stand in terms of the fraction of their cells undergoing cell division at any time—and much evidence suggests these are the cells most susceptible to cancer induction (Gofmann et al., 1967).

PRUDENT ASSUMPTIONS

Many different kinds of human cancer have been shown to be or are suspected of being caused by radiation. For other cancers, there is no evidence, one way or the other. There is also evidence that increases in radiation exposure cause proportional increases in cancer incidence. It would seem prudent to assume, and the evidence supports this assumption, that proportional increases in all forms of cancer follow increases in radiation exposure, even at doses below the radiation exposure permitted by federal regulations. There may be a threshold in some cases, but there is no firm evidence for such exceptions now.

Furthermore, the evidence we have cited here indicates that the effect of radiation can be expressed as an increase in the natural incidence of the various forms of cancer in the population that is exposed.

This is only an approximation, given the uncertainty of many of the figures we have used, and the range of estimates certainly does not rule out a variation as large as a factor of ten in the rates at which different cancers are induced. There is no reason to think that all are lower than the data now suggested, however, and the average we have chosen is probably reasonable.

A final assumption, widely shared and well-supported by evidence, is that young people are more susceptible to radiation effects than are adults.

If all these assumptions are made, the following estimates seem reasonable:

For adults:	100 Rads is the estimated doubling dose for all cancers
	1% increase in incidence rate per year follows each increase of one Rad of exposure
For those under 20 years of age:	Between 5 and 100 Rads is the estimated doubling dose
	Between 1 and 20% increase in incidence rate per year per Rad of exposure
For infants in-utero:	6 Rads is the doubling dose
	17% increase in incidence rate per year per Rad of exposure

Using these assumptions, we have calculated that the present population would experience 16,000 additional cases of cancer and leukemia each year if exposed to the extent now allowed for 30 years. If the population increases 50% over this time, the number of added cancers would increase proportionately, to 24,000. Furthermore, we would estimate that these numbers, if anything, probably underestimate the risk. For purposes of setting radiation-tolerance guidelines, one might even be advised to use lower doubling doses than those estimated above. We have not tried to estimate the effects of prenatal exposure, for instance.

In addition to man-made radiation there is, of course, natural radiation. Some persons have argued that as long as we do live in this "sea of radioactivity" we should not be concerned with adding a little more. We think it clear that although natural radiation does affect the frequency of cancer and leukemia in the same way as does "manufactured" radiation (approximately 1% increase in incidence per Rad), the 24,000 additional cases of cancer and leukemia each year constitute a significant amount of human misery which could be avoided.

We have attempted to show herein that ionizing radiation is capable of causing significant increases in incidence of most forms of cancer and leukemia. Therefore, it is of prime importance to carefully consider present guidelines that establish permissible levels of peaceful radiation and weigh the consequences of these levels against the benefits derived.

REFERENCES

Anderson, R. E., and K. Ishida, 1964. Malignant lymphoma in survivors of the atomic bomb in Hiroshima. *Ann. Internal Med.* 61:853–862.

Carroll, R. E., et al., 1964. Thyroid cancer: Cohort analyses of increasing incidence in New York State, 1941–1962. *J. Nat. Cancer Inst.* 33:277–283.

Court-Brown, W. M., and R. Doll, 1965. Mortality from cancer and other causes after radiotherapy for ankylosing spondylitis, *Brit. Med. J.* 2:1327–1332.

Gofman, J., et al., 1967. A specific common chromosomal pathway for the origin of human malignancy. UCRL-50350, November 20.

Hempelmann, L. H., 1968. Risk of thyroid neoplasms after irradiation in childhood. *Science* 160, no. 188:159–163.

Joint Committee on Atomic Energy, Hearings of 1967. Radiation exposure of uranium miners, 90th Congress, Part 2, p. 1047.

MacMahon, Brian, 1962. Epidemiologic aspects of cancer. *Cancer J. Clinicians* 19, no. 1:27–35.

MacMahon, B. and H. Hutchinson, 1964. *Rev. Acta Un. Int. Cancer* 20:1172.

Maki, H., et al., 1968. Carcinogenesis in atomic bomb survivors. Technical Report 24-68, Atomic Bomb Casualty Commission, Nov. 14, 1968.

Miller, Robert W., 1969. Delayed radiation effects in atomic bomb survivors. *Science* 156:569–574.

Pochin, E. E., 1966. Somatic risks—thyroid carcinoma. In *The evaluation of risks from radiation,* Internal Commission on Radiation Protection, *Publication 8.* Oxford: Pergamon Press.

Stewart, A., et al., 1958. Survey of childhood malignancies. *Brit. Med. J.* 1:1495–1508.

Stewart, A., and G. W. Kneale, 1968. Changes in the cancer risk associated with obstetric radiography. *Lancet* 1:104–107.

Tamplin, Arthur R., 1969. Fetal and infant mortality and the environment. *Bull. At. Sci.* 25, no. 10.

THEODORE B. TAYLOR

The Need for National and International Systems to Provide Physical Security for Fissionable Materials

As the prospect of employing more and different nuclear fuels in a wide-spread program of power production comes into focus, there is a growing need for better protection against theft than is now provided for fissionable materials used for civilian nuclear applications in the U.S. and other countries. There are several reasons for concern. Stolen fissionable materials could supply an illegal national or international market. They could be incorporated into crude but highly destructive nuclear explosives for use by extremist organizations, or they could be exploited by countries that want to make nuclear explosives but do not have direct access to required materials or optimal facilities for weapons production. At the same time, present nuclear material safeguards are less effective impediments to theft than those successfully overcome by thieves of other valuables in the past. Many situations now exist where quantities of fissionable materials sufficient for several nuclear explosives are not protected by armed guards, major physical barriers, or intrusion alarms. It is a matter of great urgency, therefore, that all nations that possess more than a few kilograms of highly enriched uranium, plutonium, or U^{233} implement systems of safeguards to prevent theft of these materials. Furthermore, the assessment of ways to meet future demands for energy must take into account the financial costs of providing such safeguards and the social costs of neglecting to do so.

All forms of energy used by man entail some degree of hazard. According to the *1971 World Almanac,* fires in the United States in 1970 killed more than 7000 people and destroyed almost 2 billion dollars worth of property. More than 5000 people died in coal mine disasters in the U.S. in the hundred years prior to 1969. Floods resulting from the collapse of the hydroelectric dam near Vaiont, Italy, killed 1800 people in 1963. Gas pipeline explosions and oil fires also take their share of human lives and extensive property damage.

Disastrous accidents are not the only hazards. Sulfur dioxide, carbon monoxide, nitrogen oxides, coal mine dust and acid drainage, spilled oil, and evaporated gasoline are all examples of byproducts of energy production that can seriously damage human health, upset the balance of nature, or make our environment distinctly unpleasant.

Nor are all the hazards unintentional. Men have diverted all types of fuel from peaceful to destructive uses. Wood converted to charcoal has been used to make millions of tons of gunpowder. Oil or natural gas are primary substances used for making high explosives employed by military forces and terrorists. Gasoline is the main ingredient in Molotov cocktails.

Nuclear energy is no exception. Although no one outside a nuclear installation has been killed as a result of a reactor accident, several such accidents have killed employees. Radionucleides from nuclear power plants and nuclear fuel reprocessing plants are potential hazards to the public if not painstakingly contained. More than a billion dollars have been spent in the U.S. alone to develop and implement safeguards against reactor accidents and other releases of radioactive materials. We are all painfully aware of the dangers associated with nuclear energy used for military purposes.

In spite of these hazards, the world's demand for energy continues to increase, and is now doubling about every fifteen years. Apparently, the benefits so far outweigh the risks. If it can be shown that the benefits of nuclear power from fission, connected to the vast potential energy in the world's reserves of uranium and thorium, greatly outweigh the risks, then I have little doubt that the rate of growth of nuclear power will continue to increase until it becomes one of the world's primary sources of energy. Before we make any long-term commitments to a particular development of nuclear energy, however, its benefits, costs, and risks must be thoroughly assessed. Otherwise, we may find that great efforts have been expended in an unacceptable direction, resulting not only in wasted time, but also in massive shortages of power.

I wish to return, therefore, to the one risk in the development of nuclear energy which concerns me here: the threat of theft of fissionable materials by individuals or groups who wish to build nuclear explosives.

The levels of physical security now applied to fissionable materials for civilian use, in the U.S. and other countries, are considerably lower

Table 1 Rough Projections of Plutonium Production Capacities, Selected Non-Nuclear-Weapon States, 1975–80

Country	Estimated Installed Nuclear Capacity 1975–1980 (Mwe)[a]	Estimated Plutonium Production Capacity 1975–1980[b] (kg. per year)
West Germany	5,000– 20,000	1,000– 4,000
Japan	5,000– 20,000	1,000– 4,000
Canada	2,500– 6,000	600– 1,500
Sweden	2,500– 4,000	500– 800
Italy	1,400– 5,000	300– 1,000
Spain	2,000– 5,000	400– 1,000
Switzerland	1,000– 3,000	200– 600
India	1,200– 2,000	300– 500
Israel	—	10–
Other	5,000– 15,000	1,000– 3,000
Rounded Total	26,000– 80,000	6,000–17,000
United States	50,000–120,000	10,000–25,000

[a]Megawatts of electric generating capacity.

[b]Based on 0.2 kg. plutonium per Mwe per year for light water reactors and a higher figure for natural uranium reactors. Note that additional time must be allowed for extraction of plutonium.

Sources: For 1975 estimates see "Power Reactors '70," *Nuclear Engineering International,* February 1970, pp. 109-30. The 1980 figures are the author's rough estimates.

than those overcome in many successful thefts of other valuables in the past.

Present civilian inventories of fissionable materials in forms suitable for use in nuclear explosives (without requiring extensive material processing facilities) are of the order of several thousand kilograms in the United States; non-U.S. inventories are probably somewhat smaller. Both are increasing rapidly (Table 1). Yet, the knowledge, materials, and facilities required to make nuclear explosives with yields at least as high as the equivalent of a few tens of tons of high explosive, given the required fissionable materials, are distributed worldwide. They could be assembled by groups of people with resources available to practically any country in the world. There is no longer any secret of the "Poor Man's Atom Bomb." While the future availability of highly enriched uranium in presently non-nuclear-weapons countries is still uncertain, the rate of reactor installation in these countries makes it fairly certain that large amounts of plutonium will be available. Gilinski (1972) explicates this as follows:

The rate of plutonium production depends on the type of reactor and the mode of operation, but, in any case, substantial production of plutonium is an unavoidable by-product of the operation of uranium-fueled power reactors. Under normal commercial operating conditions, light water reactors produce about 200 to 300 kilograms of plutonium per year for every 1000 MWe. Natural uranium power reactors typically produce plutonium at about twice this rate.

The civilian plutonium, after it is chemically separated, can be used as fuel in present-day reactors, or it can be stored for future use in advanced plutonium-fueled reactors such as "fast breeders." Our concern is that it could also be diverted to military use.

However, because fuel is usually kept in a power reactor for a relatively long time, the plutonium normally produced in these reactors contains significant amounts of the isotope plutonium-240. In military production reactors, the concentration of the contaminant plutonium-240 in the fuel rods is kept down to less than a few percent by fairly rapid fuel replacement. Typical plutonium-240 content in commercial plutonium is expected to be about 30%, compared with the usual limit of no more than a few percent plutonium-240 in plutonium meant for military use. Since for practical purposes, the plutonium-240 cannot be removed, civilian plutonium is generally not suitable for simple, predictable, efficient weapons. Nevertheless this material can be used for making relatively simple explosives that though having unpredictable yields can be guaranteed to have sufficiently high yields to be weapons of mass destruction. Whether a country might be willing to initiate a military program with unpredictable or inferior weapons cannot be easily prejudged.

Used nuclear fuel is highly radioactive and cannot be handled easily. To extract its plutonium, the fuel must be treated in special reprocessing facilities. With present technology, fuel reprocessing generally takes at least a year. Clearly, a country must have a fuel reprocessing facility before it can have independent access to plutonium produced in its reactors.

Small reprocessing facilities are not economical, but this technology is easier to acquire than enrichment technology.

But naturally, a time will come—perhaps after 1980—when the amounts of fissile materials available in the civilian economy, especially plutonium, will be so large as to dwarf reasonable military needs. At that time, the possible demands caused by a nuclear weapon program based on fissile material from the civilian economy will be relatively smaller and more easily satisfied. As noted previously, this brief summary of the projected status of plutonium flows in the nuclear industry is based to a large extent on Gilinski (1972).

The threat of nuclear retaliation, which many believe has been the cause of nuclear stability since World War II, is ineffectual when an organization with nuclear explosives need not reveal its identity. Terrorists do not always announce themselves; nor are they likely practical targets for strategic nuclear weapons.

Relatively small nuclear explosions—fizzles by standards of sophisticated nuclear weapon designers—could cause immense damage in heavily populated areas or unusually high concentrations of valuable property, such as New York's financial district during a working day, or a large nuclear power plant.

Finally, professional criminals may be motivated, simply by prospects of large profits, to steal fissionable material for sale to high bidders. Practically every highly valuable material has been traded in illegal national and international markets. It is hard to see why inadequately protected fissionable materials should be any exception.

I cannot discuss possible protective measures in any detail here, but would like to suggest a guiding principle that might be used in designing such security systems. It could be called the "Principle of Containment in Depth," because its purpose is to restrict the flow and storage of all fissionable materials within specified, authorized channels than can be physically identified. In contrast to the majority of present safeguard procedures, which are primarily designed to *detect* a theft after it has happened, measures under this principle would attempt to *prevent* the theft from occurring at all. Such detection systems would be designed to discover any flow of materials through unauthorized channels, and transmit the information directly to control centers equipped to prevent a successful completion of the theft. Also implied is the use of physical barriers and other means, including reserve units of armed guards, if necessary, to impede the theft operations long enough to bring them under control. In other words, the idea is to concentrate attention on the presence of fissionable materials where they are *not* supposed to be, rather than on precise measurements of flows and inventories of materials where they *are* supposed to be. This is not to imply that the present use of material balances to detect losses should be abandoned, but that they should not be the *primary* basis for safeguarding nuclear materials.

I can illustrate this principle by suggesting how it might be applied to plants that fabricate fuel containing plutonium. Imagine that all such plant operations are enclosed in a big box. Channels into the box for authorized input of plutonium or output of fuel elements, as well as the top, bottom, and sides of the box, are visually or instrumentally monitored continuously to detect the unauthorized removal of small quantities through the barrier. (Instruments to detect the presence of less than a gram of plutonium in less than a second, at a distance of several meters, now exist.) All penetrations not to be used for the flow of plutonium—employee entrances and exits, channels for air, water, chemicals, equipment, etc.—are also similarly monitored. In addition, heavy storage containers, metal walls, and obstructing fences are used inside the plant, to impede the unauthorized flow of materials away from authorized channels. Automatic alarm systems to detect attempts to penetrate the barriers are used as signals for rapidly deploying reserve forces of armed guards. Finally, a secondary containment system of peripheral fences and movable barriers that can be placed along vehicle routes surrounds the entire complex.

COST OF PROTECTIVE MEASURES

One can obviously postulate a level of attack that can overwhelm practically any defensive system. Nevertheless, it would be no more rational to abandon efforts to set up a security system that makes thefts much more difficult than they are now, simply because the system is not invulnerable, than to abandon the use of city police forces because they do not prevent all crime. The real question is: what level of security decreases the risk to an acceptable level, and at what cost? If the cost is greater than can be borne by the particular type of nuclear energy system that is being assessed, then we must look for alternative sources of energy.

Although I know of no detailed estimates of such costs, I believe there is a basis for optimism. Suppose that fission energy could provide us with practically unlimited, safe, pollution-free sources of power that are economically competitive with present fossil fueled sources. I find it hard to believe we would not be willing to pay as much as 10% in added costs to reduce the risk of theft to an acceptable level. We have paid more than that to provide the levels of nuclear power plant safety that we now have, via containment vessels, concrete shielding, backup emergency cooling and control systems, and quality control. By 1980, the annual costs of nuclear electric power in the United States are likely to exceed 5 billion dollars. Ten percent of this is enough to allocate several million dollars per year to each reactor (reactors are relatively easy to safeguard) and at least several million dollars per year to each major fuel fabrication or fuel reprocessing plant. Security systems supplied at such costs would be impressive deterrents to theft indeed.

ROLE OF THE INTERNATIONAL ATOMIC ENERGY AGENCY

Clearly, this is an international issue. In the past, the International Atomic Energy Agency has specifically deferred responsibility for physical safeguards to individual national governments. It would be difficult to imagine that the IAEA could ever be turned into the kind of police-like enterprise necessary to assume direct responsibility for the physical security of facilities subject to its safeguarding authority. Nevertheless, the agency could develop guidelines in the same spirit that it has developed and widely publicized guidelines for the safe handling of radioactive materials. It could also alert Member States to evident security deficiencies in national safeguards systems.

The world has little time in which to bring this problem under control. Within two or three years, plutonium will begin to be used in large quantities in light water reactors. If the material is not much more difficult to steal then than it is now, we may find that nuclear proliferation has gone considerably beyond a few major industrial powers, and that nuclear explosions in populated areas have become an added hazard to human existence.

REFERENCES

Gilinsky, Victor, 1972. Bombs and electricity. *Environment* 14, no. 7 (Sept.).
Newspaper Enterprise Association, Inc., 1970. *The 1971 world almanac.* New York, pp. 75, 804, 805, 807.

SHELDON NOVICK

Nuclear Electric Power

Despite continuing controversy, nuclear electric power in the U.S. has entered a time of extremely rapid expansion. The cheap and plentiful energy which was hoped to flow from the atom has not materialized, and some of the nuclear industry's most serious problems remain unsolved. Thermal pollution and the routine release of radioactivity are difficulties which are soluble with present technology and are in the process of solution, at least so far as the short term is concerned. But the more serious difficulties, of nuclear accidents, waste disposal, and theft of nuclear explosive materials have proved to be intractable. In fact, these problems promise to grow more difficult in coming years, if present plans for the introduction of new "fast breeder" nuclear plants are pursued.

Yet, despite its difficulties, the nuclear industry has arrived as a major commercial energy source. In January 1972, nuclear power plants built by private industry had a total generating capacity of 7,721,000 kilowatts (*Nucleonics Week*, Feb. 1972). By the end of the year, plants in operation had increased to 15,191,000 kilowatts (*Nucleonics Week*, Dec. 1972); another 46,105,000 kilowatts were under construction; and about 42,000,000 kilowatts had been ordered by utilities (*Nuclear Industry*, 1972). With orders from previous years and more orders to be added in 1973, it now seems that by 1980 there will be about 146,000,000 kilowatts of nuclear generating capacity (*Atomic Industrial Forum*, Jan. 1973).

This is an extremly rapid rate of growth, even compared to the great general expansion of electric power. Nuclear plants in operation at the beginning of 1972 represented only about 2% of all electric generating capacity, yet the commitments for 1980 represent an expansion equal to about one-third of all generating capacity in 1972 (*Steam Electric Plant Factors*). The 146,000,000 kilowatts committed for operation by

1980 represent an investment of at least 45 billion dollars for construction alone, and a comparable sum for fuel during the life of these plants.

This extraordinary investment is being made at a time when nuclear power is still faced with strenuous public controversy. The damage from thermal pollution has been a particularly visible form of damage from all electric power plants, and efforts to limit the damage from waste heat produced by nuclear plants are underway throughout the country. Because nuclear power plants release roughly 50% more hot water, per unit of electric power, than coal-burning plants, they have been identified as major sources of heat pollution. The remedy, in most cases, is the construction of cooling towers or cooling ponds which discharge the waste heat into air as well as water, and these are increasingly being added to new power plants of all kinds. The discharge of heat and water vapor to the air from cooling towers may also have adverse local effects; these must be balanced in each case against the advantages to be gained.

The problems created by the routine discharge of radioactive materials from power plants also seem to be on their way to solution. During 1972, the Atomic Energy Commission (AEC), under pressure from the courts, its own scientists, and the general public, began to move to restrict emissions from power plants. While the final results of these actions are not yet known, they will almost certainly mean more stringent regulations.

Since the 1950s there has been heated public debate over the consequences of small quantities of radiation in the environment. In earlier years, the source of the controversy was the fallout from nuclear weapons tests, which created a small but steady increase in exposure to radiation for the entire population. From that debate, a number of scientific agreements and legislative standards emerged. Radiation, it was widely agreed, could cause some damage at any level of exposure, no matter how small. This being the case, all standard-setting bodies adopted the principle that there should be no unnecessary exposure of the population to radiation and that any exposure should be balanced by a corresponding benefit. In addition, the International Commission on Radiation Protection recommended that the general population receive no more than 170 millirems of radiation exposure each year, roughly the amount already being received from natural sources. This limit applies to exposures *in addition* to those from natural sources and from medical practices. The underlying principle of risk-benefit balancing and the 170-millirem limit were adopted by the Federal Radiation Council and the AEC and now apply to all radiation exposure from the civilian nuclear power industry.

There is a certain contradiction inherent in these principles and standards. If it is agreed that radiation causes damage at any level of exposure, it makes a good deal of sense to say that any radiation exposure, since it carries a health cost, must always be justified in terms of some greater benefit. It is not entirely consistent to proceed from these assumptions to the setting of a single standard, 170 millirems, for allowable exposure from all sources (except medical practice), as federal agencies have done. Unless, of course, the separate sources of radiation exposure have been measured and balanced against their probable benefits, with the result that a sum of 170 millirems is found to be justified. One of the individuals who participated in these discussions informs us that this inconsistency was considered, but that it was felt that exposures comparable to those already received from nature could be considered insignificant under most circumstances; he added that it might be time to reexamine this assumption.

LIGHTENING RADIATION STANDARDS

The fact that no explicit balancing of costs and benefits had been done, or that at least the balancing of costs and benefits was not generally acceptable, came to light when Drs. John Gofman and Arthur Tamplin, in a series of papers beginning in 1969, publicized the implications of existing radiation standards. Adopting the theory that radiation does damage at any dose level, as had all groups engaged in setting radiation standards, Gofman and Tamplin tried to calculate what the effects on the U.S. population would be if exposures of 170 millirems per year, permitted under federal regulation, were in fact to occur. Using the experience unhappily available from the nuclear explosions at Hiroshima and Nagasaki, records of individuals exposed to X rays for medical purposes, and other data, Gofman and Tamplin estimated that "the present population would experience 16,000 additional cases of cancer and leukemia each year if exposed to the extent now allowed for 30 years."

That this is not a generally acceptable cost for nuclear power was demonstrated by the storm of protest which followed the publicizing of the Gofman-Tamplin estimates at nuclear power plant licensing proceedings. Because the two scientists were employed at the AEC's Lawrence Livermore Laboratory, their views carried a great deal of weight with the public and within the scientific community. Both the manner in which the estimates were presented, and the uses to which they were put by citizens' action groups, created a good deal of controversy. But because the Gofman-Tamplin calculations were based on widely-shared assumptions, the controversy was short-lived. In public statements, Gofman and Tamplin called for a tenfold reduction in allowable radiation exposure; at least so far as the nuclear power industry is concerned, it seems that reductions will actually be far greater. The AEC has committed itself to standards which keep radiation exposures at a level as low as is "practicable," and recent estimates indicate this may be as low as 1 or 2 millirems, a hundredfold reduction of present standards. A study, soon to be released, of the Upper Mississippi River Basin conducted by the AEC estimates that nuclear power in this region will result in an increase of about 0.2 millirem per year in average radiation exposure by the year 2000; similar regional studies in the remainder of the country are likely to produce similar results. These results would also be in agreement with a recent study conducted in the Environmental Protection Agency (*Nuclear Industry*, Nov. 1972). While these studies may be optimistic, assuming as they do the proper functioning and maintenance of nuclear equipment yet to be tested in long-term service, they do predict that present technology can keep radiation exposure levels well below 1% of present standards. In the face of such estimates, it will be surprising if the AEC does not make a comparable reduction in permissible releases when it proceeds, later this year, to specify what the lowest "practicable" emissions from nuclear power plants can be.

If the political battle over radiation standards is almost over, the scientific debate, to the extent there was such, has also ended. As the chairman of the Federal Radiation Council pointed out in a letter to Senator Edmund Muskie in 1970 (FRC, Apr. 1970):

> Gofman and Tamplin, in reaching their conclusion that the Federal Radiation Council guidelines should be reduced *now* . . . used an approach similar in principle to that used by expert advisory groups . . . in developing radiation protection standards and guidelines. This approach is based on the assumption of a direct linear and non-threshold relationship between dose and biological effect.

SHARED ASSUMPTIONS

These same assumptions were more recently used by a panel of distinguished scientists established by the National Academy of Sciences–National Research Council (NAS–NRC), which, in a report published in November 1972, explicitly confirmed many of the estimates made earlier by Gofman and Tamplin. The report reviews, point by point, the assumptions and calculational techniques they used. It concludes that exposure of the present U.S. population to 170 millirems for 30 years would result in "from roughly 3,000 to 15,000 cancer deaths annually," which is reasonably close to the Gofman-Tamplin estimate of 16,000 per year. The report also somewhat confirms more recent predictions by Joshua Lederberg and Linus Pauling that such radiation exposures would increase the mutation rate with a resulting increase in the general level of ill health. And finally, while avoiding any recommendation about change in radiation standards, the report notes that "radiation exposure averaged over the U.S. population from the developing nuclear power industry can remain less than about 1 millirem per year . . ." assuming present technology performs as hoped. This estimate lends further weight to the proposition that radiation releases can be maintained at less than 1% of present standards with "practicable" technology, and it is likely that standards will be tightened accordingly (NAS-NRC, Nov. 1972). Nuclear fuel reprocessing plants will be more difficult and expensive to control. There is only one such commercial plant now operating, but several more will be built if the industry continues to expand. The Oak Ridge National Laboratory has conducted studies which show that releases from reprocessing plants can also be held below 5 millirems at a reasonable cost, limits that will presumably be achieved if public pressure is maintained.

Attached to this prediction of low future levels of radiation release are a number of provisos, however, which point out some of the remaining difficulties in the nuclear power industry. Exposures would remain low, the report said, provided that there is: "(a) attainment and long-term maintenance of anticipated engineering performance, (b) adequate management of radioactive wastes, (c) control of sabotage and diversion of fissionable materials, (d) avoidance of catastrophic accidents" (NAS-NRC, Nov. 1972).

This is a brief summary of the problems which the nuclear industry still faces. The first point is a political one. The nuclear industry is just beginning operation on a large scale, and predictions of radiation releases and performance in general are based on estimates which stem from the nuclear industry itself and which may, judging by past performance, be optimistic. In any case, adherence to regulations is by no means assured. The present tightening of radiation standards is proceeding in response to great public pressure exerted directly and through the courts. There is some question as to whether pressure on this scale can or will be maintained for the 30 years during which plants now under construction will operate. The AEC, a fairly large agency, now has only two dozen operating plants to regulate. As time passes, plants will grow older, vigilance will relax, and the number of plants to regulate will grow very rapidly; it is not clear that present standards will necessarily be maintained.

This issue is brought into focus by a suit by six conservation groups, headed by the Conservation Society of Southern Vermont, which challenges the ability of the AEC both to regulate and to promote the nuclear industry adequately. The suit points out that the AEC is charged under federal law with regulating the nuclear industry, but the same law imposes on it the obligation to promote the development of commercial nuclear power. Many critics have pointed out that these obligations are often in conflict; the suit asks that the promotional and regulatory responsibilities of AEC be separated into independent agencies. The suit has been filed in U.S. District Court in Washington, D.C., and the AEC has asked that it be dismissed; at this writing, the court has taken no action (Strong, pers. comm.).

NUCLEAR ACCIDENTS

The ability of the AEC to regulate the industry it has fostered becomes an even more pressing question when we consider the serious problems of the industry, still far from resolution, which are summarized in the remaining points noted by the NAS-NRC report: waste management, catastrophic accidents, and theft or sabotage.

The highly radioactive wastes produced during the operation of a nuclear power plant somehow must be kept out of the biosphere forever, but a method of doing so has not yet been developed. The problem is made extremely difficult by the long period during which radioactive wastes will remain dangerous, hundreds of thousands of years. Highly radioactive wastes presently amounting to about 80 million gallons, produced largely in military programs, are now temporarily stored in large steel and concrete tanks, which require surveillance and some of which require cooling. Tanks have developed leaks, and, for this and other reasons, a more permanent means of disposal is needed. The AEC has placed heavy emphasis on a plan for eventual burial of these wastes in deep, abandoned salt mines, which, it is hoped, will remain undisturbed for thousands of years; a first attempt to initiate this program in a salt mine in Kansas failed when it was found that many abandoned wells leaked radiation. Efforts to find an area with fewer leaks have so far been unsuccessful, although the Carlsbad, New Mexico, region is likely to be the site of the next experiment. Neither the salt mine concept nor any other plan has as yet received general endorsement from the scientific community, and at present the problem must be considered unsolved. It is not clear that an entirely satisfactory solution can ever be found, as this would require complete security for the wastes over thousands of years. The difficulty is that anywhere the wastes can be placed is, by definition, accessible, and it is difficult to imagine a spot which would remain undisturbed for hundreds of thousands of years; no human institution is likely to be able to provide surveillance, or even records of waste disposal, for times extending over geologic eras.

A problem which is proving to be almost equally intractable concerns the possibility of catastrophic accidents in nuclear power plants. The underlying source of this problem is the particular design of nuclear plants in the U.S., the so-called light-water reactors, which employ enriched uranium as fuel. During the 1940s and 1950s, the U.S. built three huge plants for the separation of uranium isotopes. These gaseous diffusion plants, which represent an investment of several billion dollars, are able to separate the rare form of uranium, U-235 from the more common form, or isotope U-238. It is the rare variety, U-235, which sustains nuclear fission reactions in power plants and in nuclear weapons. The gaseous diffusion plants were built to satisfy military requirements which have long since been met. For almost twenty years, therefore, the U.S. has had a considerable capacity for separating the two isotopes of uranium, which the AEC has employed in the service of developing a civilian nuclear power industry. All of the nuclear power plants sold in the U.S. are based on designs employing uranium fuel in which the proportion of U-235 has been increased over that found naturally; this is made possible by the availability of these gaseous diffusion plants, which were built for military purposes and which represent an investment that likely would not have been made in the interests of civilian nuclear power alone. Other countries, such as Britain, which developed nuclear power without a correspondingly large military program, have chosen other reactor designs which do not require fuel with a high proportion of U-235.

Since it is the U-235 portion of the uranium which sustains the nuclear reaction, U.S. power plants using "enriched" fuel can be somewhat more compact, and therefore slightly less expensive to construct, than power plants with a lower proportion of U-235. The enrichment of U-235 content also allows the use of ordinary water to moderate the chain reaction, which otherwise might require heavy water, graphite or other more expensive materials. These advantages have given U.S. reactors a competitive edge in this country and abroad, allowing U.S. reactor manufacturers to dominate the world market for electric power. The economic advantages of enriched fuel, however, carry a serious liability. While the compactness of the fuel is desirable in terms of construction costs, it creates serious safety problems which have yet to be resolved.

In a nuclear power plant, the heat generated by nuclear fission in the fuel heats the water flowing through it; this heat is used to generate steam and, ultimately, electric power. The smaller the volume of fuel used to generate a certain quantity of heat and, hence, of electric power, the higher is the "power density" of the fuel, which can be measured as so many kilowatts per pound of fuel. Because U.S. reactors have a higher power density than those of other countries, there is more heat generated and more radioactive waste accumulated per pound of fuel.

During operation, radioactive wastes accumulate in the fuel in proportion to the power density as long as the plant is in operation. A reactor which is little used would, of course, accumulate little waste in its fuel. The wastes themselves generate a good deal of heat which, unlike the nuclear fission reactions going on in the fuel, cannot be shut off. The heat from radioactive wastes, or "decay heat," may account for 10% or more of the heat being generated in the reactor fuel and so adds to the capacity for electric power generation, but since this decay heat cannot be shut off when the plant is not operating, it creates very complex safety problems. Cooling water must continue flowing through the fuel even when the plant is not operating, or the radioactive wastes will quickly heat the fuel to the point at which it melts and undergoes chemical reactions with water and steam. This must not be allowed to happen, firstly because it would mean complete loss of a power plant worth hundreds of millions of dollars, and secondly, and more importantly, the melting of the fuel might result in the release of large quantities of radiation to the outside environment.

Such an event is usually referred to as a "catastrophic accident," and a good deal of the effort which goes into the design of nuclear plants is devoted to seeing that it does not happen. Daniel Ford and other members of the Union of Concerned Scientists have in recent months questioned the adequacy of emergency systems designed to keep cooling water flowing through a nuclear power plant under all circumstances. If the flow of cooling water should be interrupted for as short a time as one minute, the rapid heating of a reactor's fuel would cause chemical reactions and mechanical deformations that would make further cooling all but impossible, and an irreversible chain of events would begin, followed by the melting of the fuel and supporting structures. The molten fuel would melt its way downward, presumably into the earth; nuclear engineers long ago named this chain of events the "China Syndrome," as it is uncertain how far the hot mass would penerate into the earth (Novick, 1969).

HAZARDS OF DESIGN

It is important to remember that the hazards of fuel melting, while inherent in the very large U.S. plants of current design, do not arise in smaller plants or in those, like the early British reactors, which use fuel of lower enrichment. These difficulties arise only where power densities are high, as they are in all reactors currently being sold in the U.S. (Paterson, 1972). These hazards can be reduced by adding a variety of emergency cooling systems intended to go into operation should normal cooling fail, but they cannot be totally eliminated. Emergency cooling systems would have to come into operation very quickly, in a matter of seconds, under conditions which by definition would be abnormal—after an earthquake, sabotage, or some other cataclysmic event that could damage the massive normal cooling mechanism. Somewhat arbitrarily, the emergency cooling systems are designed to function following a postulated accident in which there is a double-ended break of one of the massive cooling-system pipes; in other words, it is assumed that one of these pipes is completely severed, and the cooling water, normally under high pressure, flows out of the reactor, leaving the fuel exposed. Although accidents which would have this result are not likely to be common occurrences, because of the severity of the potential consequences, the AEC has always insisted that designs take into account such a complete break in the cooling system, assuming that this is worse than the most severe accident which could occur. The possibility of still worse accidents as, for instance, the bursting of the reactor vessel or boiler, may be more real than the AEC has been willing to acknowledge; the recent hearings on emergency cooling safety have shown that some of the AEC data on vessel safety are unreliable (Kendall and Ford, 1973).

Reactor manufacturers have responded by adding to their plants a variety of spray or flooding systems which would force cooling water into the fuel mass after all normal cooling water had been lost. The adequacy of these "emergency core cooling systems" has been examined during several months of hearings before the AEC. A good deal of evidence has been adduced that emergency cooling systems, as presently designed, are uncertain at best and that present knowledge of the behavior of nuclear reactors under accident conditions is not sufficient to guarantee safe shut down after normal cooling has been lost. The AEC has already indicated the intention to reduce power levels, and hence power densities, of some older plants, but it will probably be months or years before the controversy is resolved. As with the problem of waste disposal, it is not clear that a completely satisfactory solution can be found, for the hazard of fuel melt-down is inherent in the design of current plants, and no emergency system can be absolutely guaranteed to function under any and all circumstances. Because every effort is made to mitigate the consequences, a severe accident is by definition of all foreseeable accidents an unforeseen event. Hence, emergency systems that could forestall the consequences of *any* accident may prove, in practice, impossible to design. In partial recognition of these unresolved difficulties, the AEC has continued to insist that all nuclear plants be built in unpopulated areas. There is now no provision for maintaining low population densities in the area of a plant, once built, however, during its 30-year life-span.

The importance of guaranteeing safe operation of nuclear plants under any possible circumstances stems, of course, from the truly catastrophic consequences which would follow the release of any large part of the radioactivity stored in an operating power plant. In 1957, the AEC released a study of the consequences of a hypothetical accident in the small reactors which were then planned (AEC, 1957). While this now-famous study had many serious deficiencies, it remains the basis for discussion. The study estimated that casualties following a major accident would be 46,400, including 3,400 deaths. Since modern reactors, at 1000 MWe are about five times larger than the one used as the basis for this study, and since they have much higher power densities, it can only be assumed that the potential damage from such an accident today would be still more serious. The actual damage, of course, would depend on the population downwind from the plant.

The gravity of a reactor accident, should one occur, is so great that something approaching absolute assurance of safety in nuclear plants is demanded, yet, as we have seen, this may not be possible in reactors of current design. In the Magnox reactors used in Britain, fuel with lower enrichment and power density is used, so that even with a complete breakdown, the flow of gases through the fuel would cool it sufficiently to prevent melting, without positive safety measures being employed. This is a "fail-safe" type of design (Patterson, 1972). In a U.S. reactor, by contrast, it is not difficult to postulate events that would negate all safety measures, leading to fuel melting and radiation release. For instance, bursting the steel vessel, or boiler which holds the fuel, would make all further artificial cooling extremely difficult, and would, therefore, possibly result in a catastrophic accident. The AEC thus far has not considered that this is a possible event and has not asked reactor manufacturers to take it into account in their designs.

Less dramatic events, not taken into account in present designs, could also, at least conceivably, lead to massive radiation releases. In November 1972, for instance, an anonymous letter to the AEC pointed out that the main steam line in at least one nuclear plant passed under the building housing the plant's control room; a break in the steam line might damage the building sufficiently to prevent any emergency measures being taken (Novick, 1969). The AEC moved promptly to investigate this situation, and presumably, will find measures to mitigate such an accident. The central concern is whether all such possibilities can be foreseen. Aside from damage to reactors themselves, increasing volumes of radioactive fuel and waste will be shipped about the country as the industry expands. These shipments are liable to a variety of hazards, some of which may prove unforeseen.

NUCLEAR HIJACKING

In addition to the hazards of serious accident are the possibilities of sabotage and the theft of nuclear materials, which, like the problems of waste disposal and accident prevention, cannot be completely removed, although the risks can be reduced. The current plague of aircraft hijackings has made it clear that society is highly vulnerable to determined efforts at sabotage and that these are extraordinarily difficult to prevent. The disturbing possibilities were brought into focus by a recent incident, in which the hijackers of a commercial airliner threatened to crash their captive aircraft into the nuclear facilities at Oak Ridge National Laboratory unless a ransom were delivered, and by another report that a commercial aircraft diverted to Cuba carried on board, unknown to hijackers, fissionable materials from which a nuclear explosive could have been constructed (*Washington Monthly*, Jan. 1973).

It is clearly not beyond possibility that a nuclear power plant could be held hostage for financial gain or for political purposes as aircraft now are so frequently. Power plants are so massively protected against accidents of various kinds that sabotage which released large amounts of radiation to the air or water would be difficult, but there is a real possibility that a knowledgeable individual, or an employee, might make such an attempt with the aim of obtaining a ransom, the release of political prisoners, or some other objective. It would be fairly easy to do extensive damage to a power plant, but for the general public to be endangered the reactor itself or its primary cooling system, which are massively protected, would have to be damaged from the outside. The threat of complete loss of a power plant is likely to suffice for most easily imagined saboteurs.

Also disturbing is the possibility of theft of nuclear materials for construction of a clandestine nuclear weapon. There is almost universal agreement that an individual or organization with sufficient resources to construct a nuclear weapon would have no serious difficulty in obtaining the raw materials by theft. Several incidents in which nuclear materials have been lost in transit are already on record (*Environment*, Oct. 1972), and as the nuclear industry grows, opportunities for theft increase.

The dimensions of this vexing problem can be drawn quite easily. It takes, very roughly, 5 kilograms of plutonium, or 20 kilograms of uranium 235, to construct a nuclear explosive: these are the "critical masses" of these materials under ordinary conditions. During normal operations, a nuclear power plant of this size now being built in the U.S. produces between 200 and 300 kilograms of plutonium per year (Gilinsky, 1972). Disregarding the difficulties of handling and fabricating plutonium, which are considerable, a power plant produces enough material to manufacture perhaps 50 nuclear weapons each year. The plutonium is extracted from the fuel at a reprocessing plant which is, typically, some hundreds of miles from the power plant; the plutonium is then shipped to still another installation for use in fabricating fuel or for storage. At any point in this process it is liable to theft, and current security precautions are not particularly stringent, as noted above. Since during most of its travels the plutonium is mixed with uranium and massive quantities of radioactive wastes, this may not be a matter of great concern, except with regard to the step at which plutonium is extracted from the general mixture and before it is again mixed with uranium fresh fuel. At this point, the plutonium, highly toxic and radioactive, is in solution in concentrated nitric acid. The technical difficulty of making a weapon from such starting material would be formidable, and it is reasonable to question whether some determined individual or group would not be more likely simply to make an effort to steal an existing nuclear weapon. As in the case of accidents, however, it is difficult to foresee all possible future contingencies, and so long as plutonium, in its various forms, is subject to only slight control, a real risk of diversion remains.

On February 1, 1973, the AEC issued new regulations imposing tightened safeguards on materials which could be used to build nuclear explosives. The regulations prohibit shipment of any but very small quantities of fissionable materials on domestic passenger planes and impose stricter requirements for security and periodic inventories for civilian agencies licensed to handle such materials. The changes with regard to nuclear fuel reprocessing plants are quite minor, however, and do not affect the overall situation appreciably.

As the nuclear industry grows and as literally tons of plutonium are shipped about the country annually, these problems will worsen and better measures than those now available will have to be found. Given the catastrophic nature of a single theft, it is by no means assured, once again, that thoroughly satisfactory protective measures are possible. As in other aspects of the nuclear industry, some irreducible risk of clandestine nuclear weapons diversion will be attached to the development of nuclear power. It is not entirely clear that when this risk is exposed to general public attention, the benefits of nuclear power will be generally agreed to outweigh the hazards.

BREEDING PROBLEMS

Considering the rate at which the nuclear industry is now expanding, it would be reassuring to be able to say that the problems of nuclear waste disposal, serious accident, sabotage, and theft are being diminished with equal speed. But this is not so. The current commitment to development of a new type of nuclear plant, called the "fast breeder," has potential for severely worsening all of these problems. The fast breeder would have all of the difficulties we have already described: thermal and radioactive pollution, accident and sabotage hazards, waste disposal problems, and so forth. In some cases, however, these difficulties would be severely worsened.

The attraction of the fast breeder reactor design is, once again, primarily economic. Uranium prices in recent years have been low, and explorations for new reserves have been correspondingly reduced. As the nuclear industry expands, demand for uranium will grow, and uranium prices are expected to rise dramatically in coming years. This would, of course, increase the cost of electricity from present nuclear plants.

The breeder reactor would be much more independent of uranium costs than present reactors. Although a breeder's initial fuel would be highly-enriched uranium, the bombardment of neutrons created by the nuclear reactions going on in the reactor's fuel would convert ordinary uranium 238 into plutonium, which could then be extracted and reused as fuel for the plant. More plutonium can be made than is burned in a breeder (hence the name), which allows an expanding industry to fuel itself, with much less reliance on fluctuating uranium prices.

Assuming that uranium prices do, in fact, increase in coming years, the breeder might carry an economic advantage, assuming also that it could be built and operated for not too much more than present reactors. Less mining and milling of uranium would also benefit the environment. With these advantages, however, comes a series of penalties which are by no means trivial. The most obvious is the large increase in the amounts of plutonium which would be produced and shipped about the country, greatly increasing the possibilities of theft or sabotage. Safety problems, too, would be exacerbated. For a number of technical reasons, the fuel of a breeder would be even more compact than that of present light-water reactors, and so power densities would be even higher, with all of the associated problems made more difficult in proportion. The compactness would be achieved by increasing the fraction of the fuel made up by uranium 235 (and, eventually, plutonium) in the fuel: the breeder may have fuel enriched to about 20% U-235, compared to roughly 3% in current reactors. Because of the very high power densities in the breeder, water would not be adequate to carry off the heat, and a liquid metal, sodium, will be used as the coolant (the new plants are therefore often referred to as liquid-metal cooled fast breeders, or LMFBRs). To the already difficult problem of maintaining cooling under all circumstances in conventional reactors that use water as a cooling medium, would be added the difficulty of using liquid sodium, a highly reactive, opaque, and difficult material.

Not only will the difficulties of cooling be compounded, but the hazards caused by an interruption of cooling would be even greater. Conventional reactors require the presence of a moderator to continue functioning, and, in the most common case, the cooling water itself serves as moderator. The function of the moderator is to slow the neutrons which perpetuate the nuclear chain reaction, and in conventional designs the nuclear reactions cease immediately when the water is removed. In this respect they are fail-safe—anything which damages the integrity of the reactor would lead to a loss of cooling and a shutdown of the plant. As we noted before, this would not end the production of heat by radioactive decay of waste but would at least preclude a runaway nuclear reaction.

Breeder reactors of the type now being proposed do not require the presence of a moderator; the neutrons are hence not slowed, which gives rise to the name "fast" breeder, a misleading term which implies nothing about the speed at which the reactor operates or even the rate at which the nuclear reactions proceed in its core. The "fast" refers only to the speed at which neutrons are traveling.

STOPPING A NUCLEAR RUNAWAY

Because the breeder does not require a moderator to continue operating, positive action must always be taken to shut down the plant in an emergency. Indeed, in the LMFBR, under certain circumstances, loss of the cooling medium would actually speed up the rate of nuclear reaction. In the breeder, therefore, emergency measures must always be available to shut down the plant as well as to provide continuous cooling—another drastic compounding of safety problems.

A further hazard is created by the breeder's design. In the event that cooling should fail and all, or a portion, of the fuel should melt, the result would be not just a melting through of the reactor structure, but a nuclear excursion, or runaway nuclear reaction, which would create a small explosion (not in any sense comparable to the explosion of a nuclear weapon). Such explosions have been extensively studied in theoretical terms, and several computer programs have been designed which attempt to predict the force of the explosion which would result from the melting of a fast breeder reactor's fuel. As in other efforts to predict the course of an accident, there is a good deal of contention, and one recent study seems to show that under certain circumstances an explosion could be worse than had previously been predicted—sufficiently severe, in fact, to rupture the protective structures designed to prevent radiation release in fast reactors which have been built so far (Webb, 1971).

A final, related, difficulty, is that anything which further compacts the breeder's fuel would increase the rate of nuclear reaction, and it is therefore possible that the explosion which results from a melting of the fuel would be self-enhancing (or "autocatalytic"). In general, this means that a mild explosion in one part of the fuel could, conceivably, compress the remaining fuel very rapidly, with the result that a further, and larger, explosion would follow. Given the quantities of uranium 235 or plutonium present in a fast breeder reactor, the possible consequences are disturbing (Inglis, 1969).

Since a commercial, functioning, fast breeder has yet to be built in this country, there is little we can say about how these serious difficulties are to be met. Several experimental fast reactors have been built in the U.S., and there has been one effort to operate a commercial fast breeder reactor, the Enrico Fermi Nuclear Power Plant near Detroit, but nearly all have ended in failure. The first fast breeder, the Experimental Breeder Reactor—I (EBR-I) was destroyed in 1955 while an experiment was underway; in this accident most of the reactor's fuel was melted, but no explosion, fortunately, ensued. Some years later, the Enrico Fermi reactor was constructed, and while it was planned to provide commercial electric power, a series of difficulties and mishaps prevented it from functioning properly. Finally, in 1966, another accident occurred which resulted in the melting of a small portion of its fuel, again, fortunately, without any explosion ensuing. Following this accident, efforts to bring the plant into commercial operation were abandoned, and an oil burning plant was built to replace it. After repairs from the damage were completed, the owners of the plant, principally the Detroit Edison company, tried to obtain funds to continue its operation as an experimental facility, but these efforts failed, and the plant is now permanently shut down, without ever having produced appreciable quantities of electricity or plutonium. The venture represents a dead loss of at least $150 million (*New York Times*, Nov. 30, 1972).

UNCERTAIN FUTURE

Although another experimental plant, EBR-II, has so far functioned satisfactorily, the record of fast breeders in the U.S., has not been encouraging. Britain, France, and the Soviet Union have all successfully operated experimental liquid-metal-cooled fast breeders with no serious accidents reported, but only the U.S. has so far made a public commitment for the rapid commercial development of this reactor type. The Westinghouse Corporation has been awarded the contract to design and construct a new commercial prototype fast breeder, at a cost variously estimated to range from 500 million to $1 billion. The plant will be owned and operated by the Tennessee Valley Authority (TVA) and the Commonwealth Edison company of Chicago, at a site on TVA property; $240 million is to be contributed to construction costs by private utilities, about $50 million from reactor manufacturers, and the balance, whatever it ultimately is, by the AEC, which also assumes all risk in the venture. The prototype is scheduled for operation by 1980, and the AEC expects commercial fast breeders to be widely introduced shortly thereafter. Considering the problems which have plagued this program for almost twenty years and the many serious and still unresolved questions regarding the fast breeder's safety, there is some reasonable doubt as to whether these hopes can be fulfilled or whether, in fact, they should be. The Scientists' Institute for Public Information has entered a suit in federal court asking that the AEC prepare an environmental impact statement on the program as a whole, which could serve as the focus of public debate on its merits. The suit was denied in district court and is now being appealed.

In summary, it seems clear that while some of the problems created by rapidly expanding nuclear industry are on their way to solution, others will be compounded by planned developments. The problems of waste disposal, serious accident, and the threat of theft or sabotage at nuclear facilities are not completely resolvable, and may grow more serious if the fast breeder reactor is introduced. In general, therefore, it now seems likely that the expansion of nuclear power will carry with it some serious and irreducible risks. Whether the economic benefits of the industry counterbalance these risks is a question which has never been publicly resolved.

REFERENCES

Atomic Energy Commission (AEC), 1957. *Theoretical possibilities and consequences of major accidents in large nuclear power plants.* (WASH-740), Washington, D.C., Mar.

Atomic Industrial Forum, Jan. 1973. *Info.* New York.

Environment, Apr. 1970. Letter dated Jan. 28, 1970, p. 16.

———, Oct. 1972. And now for a little diversion . . .

Gilinsky, Victor, 1972. Bomb and electricity. *Environment,* Sept.

Gofman, John W., and Arthur Tamplin, 1970. Radiation: The invisible casualties. *Environment,* Apr.

Inglis, David J., 1969. In *The careless atom.* Boston: Houghton-Mifflin, Chapter 7.

Kendall, Henry, and Daniel Ford, 1973. *Environment,* Apr.

National Academy of Sciences (NAS-NRC), 1972. The effects on population of exposure to low levels of ionizing radiation. *Report of the advisory committee on the biological effects of ionizing radiation,* Division of Medical Sciences, Washington, D.C., Nov.

New York Times, Nov. 30., 1972.

Novick, Sheldon, 1969. *The careless atom.* Boston: Houghton-Mifflin, pp. 7, 30, 39.

Nuclear Industry, Nov.–Dec., 1972.

Nucleonics Week, Feb. 24, 1972.

———, Nov., 1972, p. 4.

———, Dec. 28, 1972.

Patterson, Walt, 1972. The British atom. *Environment,* Dec.

Steam Electric Plant Factors. National Coal Policy Conference, Washington (annual).

Strong, Peter. President, Conservation Society of Southern Vermont, personal communication.

Washington Monthly, Jan. 1973.

Webb, R. E., *Some autocatalytic effects during explosive power transients in liquid metal cooled, fast breeder nuclear power reactors (LMFBR).* Doctoral dissertation, Ohio State University 1971 (available from University of Michigan, Inc., Ann Arbor, Michigan).

Weinberg, Alvin, 1973, Technology and ecology. *Bioscience*, Jan., pp. 1, 42.

Index